More Than Kin and Less Than Kind

MORE THAN KIN and LESS THAN KIND

The Evolution of Family Conflict

Douglas W. Mock

THE BELKNAP PRESS OF HARVARD UNIVERSITY PRESS
Cambridge, Massachusetts, and London, England • 2004

Printed in the United States of America

Library of Congress Cataloging-in-Publication Data

Mock, Douglas W.
More than kin and less than kind : the evolution of family conflict / Douglas W. Mock.
p. cm.
Includes bibliographical references and index.
ISBN 0-674-01285-2 (alk. paper)
1. Family—Psychological aspects. 2. Interpersonal conflict. 3. Sibling rivalry. I. Title.

HQ728.M55 2004 2003063566
306.85—dc22

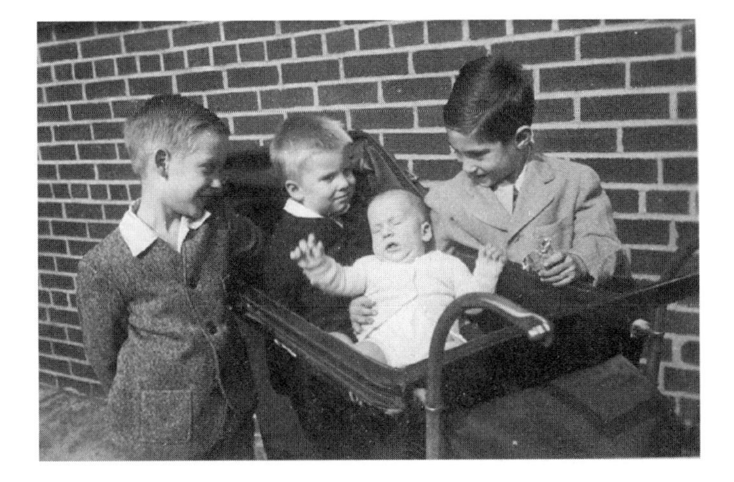

It is inconceivable that I dedicate this book to anyone other than my three older brothers, Michael (A-chick), Eric (B-chick), and John (C-chick), who taught me many useful things during our nursery days. When it became professionally important for me to appreciate the machinations of a linear dominance hierarchy, they did not have to be explained twice.

Contents

More Than Kin and Less Than Kind

Prologue

Not long ago I heard that a typical undergraduate course in introductory biology requires students to memorize more unfamiliar words than an introductory course in a foreign language. If true, this seems a terrible indictment of my chosen field. Even if untrue, it is damningly plausible. To be sure, it is easier for the cognoscenti if we agree to call that useless-looking widget of skin hanging down in the backs of our throats a uvula instead of "a useless-looking widget of throat skin." The same economy holds for prezygopophyses, lesser omentum, Loop of Henle (without which we probably would cease to honor Henle at all), and countless other pieces of jargon. This book has been written with considerable effort to minimize jargon and to define all the technical terms used. There will be no further mention of uvulas, let alone prezygopophyses, but some other terms will be necessary and some familiar words are likely to take on a new connotation. For example, *competition* within biological systems is a key concept. To illuminate this concept, let us consider two of its most extreme practitioners, namely premedical students and predators.

As an undergraduate biology major at Cornell many years ago, I honed my skills at memorization. When the need arose, I could ingest long lists of arcane terms and spit them back on demand. In truth, I was seldom sufficiently motivated to achieve perfection, but on one rare day in my sophomore year a cosmic convergence of some sort led me to a flawless performance on a practical exam in comparative anatomy. I must have expressed my joy at seeing a perfect score shining at the top of my paper, because the fellow next to me asked what my grade was. I shall never forget his response when I showed him my paper: "I hope you're happy," he snarled. "You probably just kept me out of medical school!"

Lamentably, I have no memory of my reply, which probably means my jaw dropped in silence and not on its way to saying something appropriately snippy, such as "Good news for future patients!" But the alacrity with which he translated *my* individual grade into *his* class standing and career prospects spoke volumes. Decades later, I remain impressed by premed students' obsession with grades, though I certainly recognize that their concerns are no more paranoid than Captain Yossarian's insistence that Axis antiaircraft personnel were out to kill him. The gatekeepers at medical schools place a huge premium on the GPA, and students have to play by their rules.

My classmate was very much in tune with a dynamic that ecologists call "scramble competition." The fact that I never had the slightest interest in a medical career for myself, and thus could not personally have bumped him from a hypothetical slot in med school, was irrelevant. Cornell courses were graded on a strict curve, which meant that when one student did well a burden was placed on grades-needy classmates. Parallels can be found at the racetrack, in the stock market, or anyplace where each bet or investment alters the payoffs for all other players.

Like courses graded on a curve, a family is an integrated system, with consumers on the one hand and a finite budget on the other. In many animal and plant families, the critically limited resources are nutrients available during a period of high need (such as the early growth of offspring). A brood of parasitoid wasp larvae, for example, must make do with the paralyzed caterpillar into which their mother laid her eggs. Each individual larva consumes caterpillar flesh and, in the process, reduces the total remaining for brothers and sisters. The competition lies in rapid consumption, and there may be no direct contact among the rivals whatever. They need not even be aware of one another's existence.

By contrast, an act of predation is quintessentially dyadic and, in that sense, utterly personal. If you are the diner and I am the dinner, we have an inescapable conflict of interest over how the tissues of my body are to be used. As we shall see later on, some parasitoids augment their diet of caterpillar flesh by killing and cannibalizing other consumers. One diner makes another diner its dinner, thereby reducing the number of diners and stretching the non-diner dinner. I hope that makes things clear.

Many less extreme methods exist for discouraging one's fellow consumers when the limited resource is a bit detached (say, a morsel of food that both parties want or need). It will pay to fight over access to that resource if the value gained from eating it exceeds the costs of combat. Such a process is called "interference competition" (or "contest competition"), because the fate of the resource is determined by direct physical intervention, where might makes right, rather than by speed of consumption. It's much more dyadic and personal than a scramble. Nestling egrets provide an example of interference competition among siblings: a slightly older chick may peck and intimidate its smaller nestmate, thereby discouraging it from competing for the food delivered by the parents. Beating a rival into submission solves the primary problem of there being food enough for just one. Whether the loser dies (of injuries and/or starvation) can be regarded as a separate matter.

The common denominator in scramble and interference competition is the critical shortage of resources, a fundamental mismatch wherein demand exceeds supply. Much of this book will be about food, but almost anything desirable (such as admission to med school) can spark rivalry. In the Book of Genesis, we find Jacob noting his older brother Esau's hunger and trading him stew for his birthright. The situation is different in the natural world. Among egrets, eagles, and boobies, the eldest sibling does not have to bargain with its nestmates for a meal. Instead, it typically enjoys advantages of size, strength, and experience that enable it to seize a disproportionate share of the family's food supply. And long before humans formalized traditions of primogeniture that stipulated the passing of the family property to the eldest son, parental territories tended to be inherited by the dominant offspring—usually the eldest. We humans fancy ourselves a cut above the rest of the animal kingdom, yet the histories of our royal families are replete with murdered sultans and crown princes locked in towers. For example, in A.D. 978 Edward II (the Martyr) was fatally stabbed as a teenager so that his half-brother Aethelred (the Unready) could ascend to the throne of England. A generation later, Edmund, Aethelred's eldest son, was murdered by his younger brother Knut for the same reason. So, is our story any loftier than that of honeybee proto-queens stinging rival sisters to death for the chance to become queen? From lowly dung beetles to the courts of human royalty, even the simplest families are anything but simple.

1

In a Family Way

The elephant is reckoned to be the slowest breeder of all known animals, and I have taken some pains to estimate its probable minimum rate of natural increase; it will be safest to assume that it begins breeding when thirty years old, and goes on breeding till ninety years old, bringing forth six young in the interval, and surviving till one hundred years old; if this be so, after a period of from 740 to 750 years there would be nearly nineteen million elephants alive, descended from the first pair.

—Charles Darwin

A yearling Galápagos fur seal shuffles across the warm black lava rock to bask by his mother's side. His newborn baby brother snores softly in front of her. What a tranquil domestic scene. Suddenly, the yearling plunges forward, snatches the infant by the throat, and races off with Mom in hot pursuit. The baby is tossed into the air several times before Mom catches up and seizes the victim by his rear flippers. With the parent pulling at one end and the yearling tugging at the other, the pup dies quickly. What on earth is going on here? When Solomon offered to divide a baby everyone thought him a sly genius, but how are we to understand this kind of activity in nature?

As it turns out, there are fatal battles in nests of newly hatched birds, in hyena and fox dens, and in roadside ditches teeming with cannibalistic tadpoles. Birds do it, bees do it. Families have cozy aspects, to be sure, but also decidedly harsh ones. Whenever two or more offspring depend on resources provided by their parents—whether those groceries arrive as periodic deliveries or were previously stockpiled—the potential for demand to surpass supply exists and sibling rivalry kicks in automatically. If resource shortages are brief or mild, the rivalry may be negligible; otherwise it can be fatal. In the case of the fur seals, the mother appears to hold no grudge and allows the yearling to continue nursing for another year, maybe two.

In this book, the logic of natural selection will be used to explore many complexities of family relationships. Most of the focus will be on a very elementary "mom plus a couple of kids" type of family structure (participatory dads being rather exotic accessories), but even this will provide us with much to consider. We will spend relatively little time on human families. A framework that helps us understand the social dramas within families as diverse as gulls, seals, toads, and oak trees has the potential for helping us see our own familial dynamics as a special case of a ubiquitous phenomenon.

A family is intrinsically different from a scout troop or a bird flock in an obvious way: its members are much more genetically similar. This essential feature allows the operation of an evolutionary process known as "kin selection," or more formally, "inclusive fitness theory."[1] This is a simple extension of Darwin's original concept of natural selection. The initial Darwinian view focused overwhelmingly on how an individual's own reproductive success contributes to the future distribution of traits within its population. Now that we know about the existence of genes, as Darwin did not, we recognize that certain traits stick around because their underlying genes increase in the population's gene pool. Basically, individuals that produce more offspring than average tend to leave more copies of their particular genes than their less prolific counterparts. It follows that the traits affected by those successful genes will be more common.

Kin selection is a second, less direct way for an individual to bequeath copies of its genes, by enhancing the reproductive success of close blood relatives. In brief, a trait that seems downright altruistic (benefiting someone other than the owner of the trait) can spread through populations because of the advantages that accrue through the reproduction of the trait-owner's brothers, sisters, or cousins. Much of the biology of the family rests on a very simple mathematical relationship called Hamilton's rule, which hinges on the degree of relatedness. (We will get to that in Chapter 2.)

Most aspects of behavior and ecology do not impinge directly on kin at all, but are simply between the individual and its environment. If a butterfly lands on a flower and consumes some nectar, that butterfly's body gains the energy; its siblings and cousins are unaffected. But some behaviors can have such vital effects on relatives that they probably have helped shape extravagant forms of cooperation and self-sacrifice. For example, suppose we substitute a worker honeybee for our butterfly. The bee consumes the nectar and then characteristically flies straight back to the hive, where she hands out free samples of her meal, along with detailed code telling her sisters how to find the distant flower patch. That is, she hands out valuable information. As I discuss in Chapter 12, honeybees also carry out suicide missions in defense of the hive. Their social lives are complicated. For the moment, the important thing is to note that kin selection can provide bonds between individuals, promoting certain altruistic behaviors that we humans prize as fine and noble, but that would have been exceedingly unlikely to evolve without this indirect dividend. If we celebrate aspects of the natural world that promote cooperation and social harmony, then gene-sharing among blood relatives surely counts as the Good Side of the Force.

Against this backdrop of opportunity for cooperation to evolve, a powerful counter-force lurks. Because a family often lives within a discrete area, with all its members relying on an exhaustible supply of resources, there is a strong potential for severe local competition if and when key resources run short. Consider a nestful of freshly hatched baby house sparrows, one of the most abundant birds in the world: in house sparrow families both parents collect food for the youngsters, but the chicks grow so rapidly that they pass from naked and blind infants roughly the size of a garden pea to adult-sized, fully feathered, and strong-flying adolescents in just two weeks. This requires maybe five thousand parental trips back and forth between the nest and whatever patches of insect prey the folks manage to locate. And during these two weeks a great many unpredictable things can go wrong. A cold snap or a few days of sustained rain may depress insect activity, making the birds' prey less conspicuous and thus harder to find. Cold weather may also force one parent to stay home and provide radiant body warmth for the chicks. With reduced grocery delivery and a greater portion of the nestlings' food energy diverted from growth to keeping warm, the problems worsen and sibling rivalry is exacerbated. Similar squeezes can result from any other factors that depress the food supply, imperil the parents (as when another pair tries to usurp their nesting cavity, raising their defense costs), or increase the metabolic costs of the nestlings (say, the presence of a pathogen or parasite).

In fact the predicament is even more sinister than this. Sibling rivalry does not arise only because of unforeseeable downturns in family fortunes; it also stems from adults' penchant for creating more progeny than they are ordinarily able to support. In many animal and plant species parents produce a surplus of offspring, a brood of youngsters that might, in principle, all succeed if everything worked out perfectly. But things seldom work out perfectly, so parents must, at some level, be *counting on* a culling process to eliminate some of those hungry mouths when reality fails to match perfection.

When I say parents are counting on losing some of their offspring, of course, I do not mean they are engaging in a human-like intellectual process, actively thinking about the problem of scarce resources and deriving their best solution. The parents discussed in this book range from trees to fishes to whooping cranes, none of which can be assumed to understand algebra and probability theory. They need understand nothing at all; natural selection simply favors the genes of the ones that make the best of breeding.

To us humans, though, such overproduction may seem, at first glance, like a feckless, profligate, and even sadistic thing for parents to do, so we must take a close look at why it has evolved repeatedly. Whatever the reasons, the tandem forces of parental overproduction and environmental instability have teamed up to create countless animal and plant systems in which the young compete ecologically, often to fatal extremes, with their brothers and sisters. In evolutionary terms, this means that a genetic allele that helps shape a self-promoting social trait may be favored by natural selection if ecological conditions are sufficiently severe (for more on alleles, see Box). If the crunch is really nasty, a gene can spread that promotes the survival of the Self's body even at the possible cost of copies housed in the bodies of close kin. For an eaglet, this can mean sacrificing a nestmate rather than risking starvation. In short, the economics of family life can surpass the point of balance where kin selection can keep selfishness in check.

How Genes Shuffle

We humans have twenty-three pairs of chromosomes in every cell, thus many millions of complete sets of instructions on how to build the body. Because our chromosomes (and those of most animals and many plants) come in matching pairs, we are said to be "diploid." The two chromosomes of a pair are homologues, meaning that a site on one of them that codes for a particular trait (say, a protein that affects eye color) will have an exactly corresponding site on the companion chromosome. These two versions, or alleles, of the same gene may be either identical or different. If they are identical, as in the genetic basis for blue eye color in humans, the gene is said to be "homozygous" in that individual.

Some alleles are dominant, others recessive. When one allele of a pair is dominant and the other is recessive, the dominant one determines how the trait develops. When referred to in writing, dominant alleles are capitalized; recessive alleles are lowercased. Thus, because blue eye color is recessive in humans, the diploid genotype for blue eyes is written *bb.* If the alleles differ, the gene is "heterozygous" at that locus, the genotype is written *Bb,* and because the allele for brown eye color is dominant, the person's eyes are brown.

The fertilized egg or zygote gets half its genetic material from its mother (in the ovum) and half from its father (in the sperm). When parents create eggs or sperm, their diploid cells undergo the special type of division called meiosis, wherein the homologous pairs line up at the cell's fifty-yard line and then migrate in opposite directions. The sex cells, thus, are haploid, containing only one of each original chromosome pair. The subsequent sexual act, uniting ovum and sperm, restores the diploid condition for the development of a new embryo.

Every cell in a diploid body except sex cells is essentially a copy of that first zygote. After fertilization, the new diploid cell starts cranking out replicas of itself through the standard method of cell division, mitosis. Mitosis differs from meiosis in that the cell copies its chromosomes first before splitting into two daughter cells, which are genetically identical to each other and to the original (diploid) cell. Mitosis is occurring all over our bodies every minute (which is why we can afford to shed dead skin cells, for example), but meiosis is restricted to the gonads.

Thus we have two very powerful and opposing forces, each helping shape the social relations among family members. Sometimes the balance tilts toward ecological plenty and kin interact with respect and generosity; at other times family members are one another's chief rivals, depriving and even executing one another. Viewed this way, the basic family unit can be seen as a crucible for testing the upper evolutionary limits of selfishness, because within families the closest genetic relatives are routinely locked in mortal struggles over critically limited resources. The battle lines are drawn.

A great diversity of family interactions results from the variation of these key factors across plant and animal groups. In some species, for example, all the nestmates are full siblings—that is, they have both genetic parents in common—and thus share a broader zone of evolutionary interests than half-siblings would. Here the glue of kin selection is pretty strong, because the chance that a given sib carries a copy of any particular gene is precisely one in two (for reasons to be explained in Chapter 2). In far more species, nestmates are likely to be half-siblings—sharing only one parent, usually the mother—so the chance that a sib carries a copy of a given gene is just one in four, and the glue is proportionally weaker. In yet other species, nestmates may not be close genetic relatives at all (consider tadpoles in a large pond where many female frogs have deposited eggs, or avian brood parasites, such as cuckoos, that sneak their eggs into other species' nests). Here, we expect no glue.

Parents of different animal and plant groups also vary widely in the degree to which they overproduce progeny in the first place: the number of expendable offspring may range from zero (with no overproduction), to one extra baby per cycle (as in many birds), and on up to tens of thousands (as in peach trees). In some cases, parents may participate actively in the pruning back of family size; in others, that task is left to the stronger siblings.

Family life is further complicated by environmental factors that shift unpredictably. Weather is a classically hard to predict factor, but there are many others. For example, most plants build their flowers without being able to anticipate how much pollen (which contains sperm) will be delivered by wind or animal vectors to the female flower parts, the ovules. And once the pollen arrives and plant offspring (fruits, seeds) get started, the parent's reproductive success will be shaped by an unknowable amount of damage imposed by fruit- and seed-eating insects and other animals, not to mention variation in soil conditions (nitrogen, minerals, moisture, trace elements), temperature, and so on. Circumstances may turn out to be nearly perfect; more likely many of the offspring will succumb to such unpredictable dangers. Because the upper limit on family size can be set only once per breeding cycle, at the very outset, it has to be high enough to allow for this attrition.

Finally, the balance between cooperation and strife within families can also be molded by various properties of the limiting resources themselves. Most commonly, it is food that runs short, and the details of how individual offspring gain access to food can matter a lot. A mammalian mother, for example, may be unable to provide enough milk for her entire litter, but in most cases she has more nipples than progeny. How might one littermate out-suck the competition? The answer may hinge in part on variations in the timing of milk let-down (the moment when milk is expressed from the alveolar glands into the nipple and thus becomes available). In some mammals, the milk collects for hours in large storage organs called cisterns (a bovine udder comes to mind), but in rats there are frequent and sporadic ejections of tiny amounts of milk (perhaps three to twelve times per hour, with typical volumes of half a cubic centimeter or less).[2] The types of early sibling competition that evolve in different mammals are thus likely to center on preexisting biological features that might confer modest advantages in suckling. There is, for instance, the phenomenon of rapid nipple-switching in rat and mouse pups. Because rodent mothers have roughly twice as many teats as they have babies, a pup that gulps down the minuscule serving of milk from one nipple may be able to get a second helping at another nipple. The key to success in such a rivalry clearly is drinking speed, so this is a scramble competition (rather like an Easter Egg hunt for scurrying toddlers).

Very different sets of rules must apply to birds, sharks, burying beetles, and olives, none of which have lactating parents. Some of their competitions do resemble scrambles. For example, nestling songbirds typically call and open their mouths wide when a parent arrives with food. Nobody is quite sure what these baby signals mean, but one school of thought is that the hungry offspring are engaged in a scramble competition to get closest to the food, which the undiscerning parent drops into the first available throat. In an open cup nest, the next delivery might come from any direction, so the chicks must be ready to orient and hustle when the parent approaches.

By contrast, if the food always arrives from an entirely predictable direction (as in a kingfisher's burrow nest, where the parents deliver fish through the only entrance), stronger siblings may be able to monopolize a key position by means of aggression and physical domination. Interference competition thus becomes cost-effective.

This chapter opened with a graphic example of an aggressive sib-killing for a less concrete prize, the chance to keep nursing. As it happens, such overt executions are not particularly common in Galápagos fur seals, but the neonate in a mixed-age sibship usually dies anyway from an odd sort of scramble competition. Fritz Trillmich has found that about half of all mothers nursing a one- or two-year-old pup give birth to new babies, and that nearly all of those newborns (80 percent) die within their first month—*of starvation.*[3] He suspects that the immediate problem for the neonate is that its mother's lactation system is already tuned to the stronger suckling of the older sibling, so that the newborn's feebler attempts to feed do not stimulate the release of milk. If so, most of the neonate deaths can probably be attributed to aspects of the mother's physiology. Because the mother has a central role, we must begin to entertain the possibility of her complicity.

The word "complicity" may seem to imply motivation and comprehension well beyond that intended. In an evolutionary framework, all it suggests is that the mother's overall genetic interests may be as well served by having the newborn pup die young as by having it live. Let us imagine that a given neonate has a fairly small chance of surviving early-life hazards, but that its chances get better as it passes each hurdle. At birth, then, its future may be worth rather little to its mother, but if it is still alive at the end of its first year, it carries much greater value as a potential generator of grandchildren. Allowing it to continue nursing may improve its prospects substantially. In principle, we could plug some values into those probabilities and explore whether a mother's

A mother Galápagos fur seal barks at her yearling to keep it away from its newborn sibling. If the new pup dies, the yearling will get to nurse for another year or two. (Photo: Fritz Trillmich.)

fitness is better served by investing in her healthy yearling or by terminating that offspring's support in favor of feeding the newborn. Alas, we do not know what those values are. In any case, my point is merely that the yearling *may* be the mother's better bet. If so, then losing the new pup may actually be in her own interest. That is what I mean by suggesting the possibility of maternal complicity.

Just like my saying that parents may count on losing some of their offspring, asserting that the mother seal is complicit in the death of her newborn is not meant to imply conscious culpability. When animals (and plants, for that matter) do things that strike us "clever," we should not assume they are exercising high levels of cognitive sophistication.

For example, the birds I have studied most carefully, herons and egrets, generally obtain their food by wading in shallow water and striking at fish swimming by. They are really good at this task, not making many inaccurate strikes. But the task is not a simple one. Just as, seen from above, a rowboat's oar appears to bend upward when it enters the water, the fish a heron sees is likely to be lower in the water than it looks. From physics we know that this is because water molecules are denser than air molecules, causing light to refract, but don't try explaining that to a heron or any of the other birds that catch fish in similar ways (kingfishers, terns, pelicans, boobies, and so on). So, obviously, there have to be non-analytical ways of getting the job done. Practice may help: yearling pelicans tend to dive vertically (and thus don't have to correct for refraction), but older birds can strike at flatter angles, a skill that helps them catch more fish.[4] We do not know just how they do it, but we know they don't use computers or advanced mathematics.

Getting back to the mother Galápagos fur seal and her yearling, their tug-of-war over the newborn certainly looks like a conflict and may truly represent an evolutionary conflict of interest. The scientific challenge is to keep an open mind while testing the possibilities.

2

The Problem with Sex

I would risk my life for two brothers or eight cousins.

—attributed to J. B. S. Haldane

In the simplest sense, sibling rivalry occurs because of sex, a paradoxically divisive force in nature. The paradox is in our usage of the word "sex," which means many things. In its essence, sex is one mode of reproduction. We humans usually regard it as the only procreative game in town because it is the one we use. Accordingly, we apply the word "sex" to all manner of things associated with mutual attraction, mating, and lascivious thoughts. But if we put that mind-set aside for the moment and focus on what sex actually accomplishes, a different perspective emerges, one with quite different implications for how organisms interact. In animals and plants that reproduce asexually (for example, by cloning identical copies of themselves), near-perfect social harmony reigns. By contrast, sex causes a great shuffling of the DNA cards with each reproductive episode. This is what causes each of us humans to be unique genetically, just as snowflakes are said to be unique structurally.

This effect of sex is critical to our discussion of siblings. With the exception of identical twins (which result when the first two daughter cells produced by the division of a newly fertilized egg happen to separate physically), siblings are as genetically similar/dissimilar to one another as each is to its parents and as each will be to its own offspring. The degree to which even siblings are *dissimilar* reflects their potential for evolutionary conflicts of interest.

To understand this clearly requires that one think about natural selection from the vantage of individual genes. Ultimately, natural selection is a competition among different alleles belonging to the same genetic locus (that is, different versions of the same "gene"). An allele that helps its body live longer or, more accurately, helps it breed more frequently and more successfully, tends to spread, so the physical traits it promotes become more common in the population of future bodies. We cannot observe these allelic dramas directly, but we can see their outcomes, including all the visible traits of anatomy, physiology, and behavior that are the tangible properties of each whole body. There is a crucial connection between the DNA composition (the invisible genotype) and these traits (called phenotypes).

One of the hardest things for most people to grasp in all this concerns the cumulative effects of tiny changes over huge numbers of generations. Natural selection moves in mysterious ways, through slow seepage. Small, local shifts in gene frequencies occasionally spread through whole populations and species. More often a new mutation or genetic recombination just fizzles out. The process is inexorable. It slowly and automatically tends to reward relative success, like a long-forgotten portfolio of Standard Oil shares. Even a slight genetic difference may produce a body whose performance is quite different, and it is the whole body's lifetime tally of successful reproduction that drives the alleles' replication rates. Alleles that promote greater bodily survival and reproductive success tend to spread, and thus come to replace rival alleles that predispose their bodies to make mistakes.

One type of mistake that bodies sometimes make is to mate with close kin. This is not inevitably harmful, but in many cases it leads to a problem known as "inbreeding depression." This refers to a situation in which two harmful recessive alleles happen to land together in the same zygote, so that neither is masked by a dominant allelic partner, and start building an offspring. Many familiar examples exist in human genetics, perhaps the most famous being sickle-cell anemia. When two heterozygous carriers (Xx) mate, a quarter of their offspring will be homozygous-recessive (xx) and will die from having the sickle-cell trait (their red blood cells are crescent-shaped, hence prone to jamming in capillaries). But the recessive allele persists in populations because in heterozygous (Xx) carriers it actually confers a major physiological advantage—resistance to malaria. On balance, it breaks even in regions where malaria is common.

Many other genes have alleles that are retained at low frequencies in populations for the simple reason that they seldom get the opportunity to do much harm, hence are not sufficiently penalized to be eradicated by natural selection. Imagine an allele that is both recessive and so rare that only one individual in a thousand carries it. The few carriers are nearly always heterozygous (say, *Gg*) because *g* is so scarce. The probability that two such carriers will happen to mate with each other by random chance alone is $(1/_{1,000})^2$ or one in a million. But if close genetic kin mate with each other, this probability leaps upward dramatically. A rare allele in you has a one-in-eight chance of existing as an identical copy in your first cousin. And if you carry several such rare but deleterious recessive alleles in your cells, cousin-cousin mating greatly raises the probability of having severely disadvantaged offspring.

The courtship routines of many animals include sensory exchanges that reduce the likelihood of mating between close kin. Even before the courtship stage is reached, inbreeding problems are often precluded by the tendency for members of one sex to disperse while the other sex remains close to home. In mammals, for example, the typical pattern is for male offspring to move away prior to puberty; in most birds, the exodus is by females.

Avoidance of inbreeding has two major implications for the genetic structure of families. First, because outbreeding parents are genetically dissimilar to each other, any cooperation between them after fertilization, including shared parental duties, must derive from a different sort of glue than kin selection (see Chapter 12). Second, concurrent siblings are only an even-money bet to share copies of a given rare allele, and the odds are that high only if both nestmates were sired by the same male. Our next step will be to explain why that is so.

The chance that a baby elephant (or a member of any other diploid species) contains a copy of any allele that its father carried is exactly one in two, or .50. This is the coin-flip probability over which allele's chromosome migrated into the one successful sperm that happened to fertilize the baby's egg. That chance, formally called the "coefficient of relatedness" and conventionally symbolized as *r*, is fundamental to much of what follows.

We can extend this coefficient of relatedness to other kin relationships in a straightforward way. Let's say that I happen to have a rare mutant allele that we'll call *g′*, located on a tiny bit of chromosome #10 inside every diploid cell of my body. The prime mark is tacked onto that allele purely to brand this peculiar version of gene *g*, so we can recognize our mutant each time we see it (in field research with birds, we use metal leg bands with tiny numbers engraved in them, colored plastic leg bands, small spots of feather dye, and so on for similar ID purposes). This particular mutant allele *g′* is so rare in the population gene pool (say one in a million humans have it) that there is virtually no chance that its counterpart on my other homologue of chromosome #10 also carries that same mutant. So I have just the one copy on just one chromosome: for this locus, I am heterozygous, say *gg′*. What is the chance that my sister Annabelle is also a carrier?

If you answered 10^{-6} (one in a million), take a deep breath. Assuming only that Annabelle and I are full siblings, the chance is actually one in two. First of all, if I am a carrier I *have* to have received the allele in either the egg or the sperm that got me started, which means that one of our parents is definitely a carrier. That tremendously increases Annabelle's odds of getting a copy from the same source. We do not need to know which parent carries the mutant allele, because a heterozygous carrier will pass along copies in precisely 50 percent of the (millions of) sperm or (thousands of) eggs he or she produces. Each time a meiotic cell division occurs in a parental gonad, one chromosome goes right and its homologue goes left, guaranteeing equal numbers of copies of *g* and *g′*. So half of my full siblings will be carriers and half will not.

If, though, Annabelle is only my half-sibling (say we have the same mom but two different dads), then it obviously makes a great difference whether my copy of the mutant allele came from our mother or my father. As a carrier, I must have received it from one of my parents (a fifty-fifty chance for either route), but the two routes of inheritance are no longer equivalent in terms of Annabelle's probability. The one-in-two chance that I got mine from my dad means there is virtually no chance (one in a million being roughly the same as zero) that Annabelle's father also had it. Because Annabelle and I do not share a father, our total relatedness must come from our mother. On the maternal side of the calculation, things remain unchanged. The one-in-two chance that I got my rare allele from our mom leads inescapably to the one-in-two chance that the ovum producing Annabelle also carried a copy. Because both of these contingencies must have occurred for my sib to have such a copy, the resulting probability is $\frac{1}{2} \times \frac{1}{2} = \frac{1}{4}$. Sure enough, $r = .25$ between half-siblings.

This logic of relatedness provides a relatively simple framework for exploring the complexity of family social bonds. For example, suppose that a mother not only avoids incest (and the penalties of inbreeding depression) by choosing an unrelated male partner, but actually chooses *two* such partners and mates with both of them. What is the probability that her offspring will be full sibs rather than half-sibs? For the moment, assume that both males have an equal chance (the female copulates equally often with each of them, both introduce comparable numbers of sperm into her reproductive tract, neither has sperm of superior motility, and so on). Let's say I am one of the offspring and consider my relatedness to my trusty but hypothetical sibling, Annabelle.

In these circumstances, the chance that a rare mutant allele in each of my cells also exists in Annabelle's cells is lower than it would be if our mother had mated with only one man. We can use first principles to see that it falls, in fact, from .50 to .375. On the maternal side, nothing has changed: if I got the allele from our mom, then there's a coin-flip's chance that Annabelle did too. The maternal contribution to r remains at .25. But the paternal side of the calculation clearly has been affected. Intuitively we sense that the paternal contribution must have been reduced by half, and that is true. There's a fifty-fifty chance that I received the rare allele in a sperm, in which case my genetic father is a carrier, and he has a fifty-fifty chance of having passed it along to Annabelle, but only if he is the same fellow who fertilized the egg that created Annabelle. The paternal contribution to r may still be .25 (if she and I do happen to have the same father), but it is just as likely to be zero (if we do not). So the paternal contribution falls, on average, to half its original value (from .25 to .125), while the maternal contribution remains at .25, producing the new total r of .375.

Using similar bits of basic probability logic, we can see that a great number of factors have the potential to affect r. The coefficient climbs upward if the assumptions concerning male equivalency are relaxed; for example, if one of the males provides three times as many sperm, *r* between the two randomly chosen siblings rises from .375 to just below .44. And there is a vast diversity of features in the mating systems of various animals and plants that tweak sib-sib *r*.

Consider hay fever. Every spring and summer, millions of people sneeze themselves dizzy as male plants ejaculate into the air, so to speak, and depend on the winds of chance to deliver their pollen to the appropriate female flowers. The males do this for the obvious reason that they cannot move about and find mates for themselves, behavioral attributes that help keep animal sex somewhat less random than the plant variety. Instead, male flowers must produce astonishing quantities of pollen and rely on good fortune. There is no guarantee that such a practice will work for a particular male. One effect of all this is that the fertilization of ova in the female flowers of wind-pollinated species is quite haphazard. The odds must be exceedingly low that two adjacent fruits on a given maternal plant were sired by the same male plant residing somewhere upwind. As a result, the paternal contribution to *r* is likely to be vanishingly small, so a given female's progeny are maternal half-sibs and, for a diploid species, *r* is essentially .25.

For bee-pollinated plants, mating is a bit tidier and the implications for sib-sib relatedness may differ. The pollen transferred to two flowers on the same plant is somewhat more likely to have come from the same nectar-harvesting bee. But even one bee is likely to deposit a mix of pollen loads from the numerous flowers it has visited. So, again assuming a diploid species, sib-sib *r* for adjacent fruits may be anywhere between .25 and .50.

So far, I have restricted the discussion to organisms in which the maternal contribution to sibling relatedness is clear. This assumption is comfortable for us humans because, as mammals, we have lengthy pregnancy and lactation periods that tend to promote clarity in the mother-offspring relationship. About the only ways a human baby can be assigned to the wrong mother involve kidnapping, adoption, or criminal negligence in the maternity ward. In contrast, we are fully aware of the potential for misassigned paternity, hence the saying "Mama's baby, Papa's maybe."

But there are other organisms that reverse the mammalian model, and for them the maternal contribution to relatedness between two siblings is less clear-cut. Consider sea horses and pipefishes. In these tiny creatures, males have ventral pouches in which multiple females deposit eggs before the male adds his sperm. Fertilization thus occurs inside the male's body (we might say he gets pregnant), and the developing embryos are either full siblings (sharing a mother also) or paternal half-siblings.[1]

Now that we understand the coefficient of relatedness, we are ready to consider the theoretical heart of family social evolution: Hamilton's rule. It is named for the British biologist W. D. Hamilton, who introduced it in a pair of landmark 1964 papers derived from his Ph.D. thesis at the University of London. The epigraph to this chapter, a quip often attributed to the geneticist J. B. S. Haldane, expresses the essence of Hamilton's rule. Regardless of whether Haldane actually said it, the point is immediately clear.[2] All else being equal, two full siblings should be the evolutionary equivalent of one Self. An allele that predisposes its body to take a serious risk while saving two of the body's brothers can, on average, come out ahead because it is likely to be saving a copy of itself without necessarily sacrificing the body in which it already resides. The "eight cousins" option fits perfectly, too, because a diploid individual's coefficient of relatedness to its first cousin is exactly one-eighth (assuming no hanky-panky by the grandparents).

An allele that somehow promotes an altruistic social act can spread by natural selection if the aid that it renders to the fitness of a recipient outweighs the cost that it incurs to its own fitness, once that benefit has been devalued by the closeness of the genetic relationship between the two parties. More formally, Hamilton's rule can be written as $br - c > 0$, where b is the beneficial impact on the Darwinian fitness of the recipient of any social act, c is the Darwinian cost to the individual performing the social act, and r is the coefficient of relatedness between the two parties. The resulting inequality specifies the conditions under which natural selection should favor acts of apparent altruism.

One instructive way of looking at this point is, once again, to take the perspective of the allele itself, as if it were conscious and keeping score.

The most straightforward method for an allele to replicate itself is to promote successful reproduction by its body; this is sometimes called the direct component of fitness. But as we have just seen, the allele may also follow an indirect route, providing cost-effective assistance to other bodies likely to be carrying copies. In a diploid, sexual species like *Homo sapiens,* the chance that a full sibling is a carrier is exactly the same, on average, as the chance that an offspring is a carrier. The sib-sib *r* is an *average* of .50 because of the summed maternal and paternal contributions, while the parent-offspring *r* is *precisely* .50 because of the meiotic coin-flip.

Hamilton's insight opened our eyes to the indirect component of the overall Darwinian fitness (which he called "inclusive" fitness) that an organism achieves by contributing to successful reproduction by its lateral kin. His rule offered a potential explanation for what appeared to be altruistic behavior, which had been extremely difficult to reconcile with Darwinian orthodoxy. It seemed possible that closer scrutiny of reported cases of altruistic sacrifice might reveal that they involved only modest degrees of self-abnegation, while handsomely rewarding the actor's close genetic relatives. It took a few years for biologists to understand and appreciate Hamilton's rule, but by the 1970s cadres of field researchers were rushing out to test these possibilities on such phenomena as mammalian alarm-calling (wherein the caller seems to risk attracting a predator's attention while the valuable information regarding imminent danger goes to its neighbors) and "helpers at the nest" (cases in which reproductively mature animals assist others in raising young, rather than breeding themselves).

For example, Paul Sherman, studying Belding's ground squirrels, showed that alarm calls do increase the caller's risk of becoming the predator's victim, so the cost of calling may be high. The next question was who received the benefits. Because males of this species, like most other mammals, tend to disperse before breeding, the closest genetic relationships in a given area exist among females. Sure enough, the callers turned out to be predominantly adult females, with mature females that were known to have relatives living nearby calling most of all. In some cases, these neighboring kin were cousins, aunts, and nieces, not merely offspring of the calling female.[3]

Similarly, a great many long-term studies of helping behavior in various birds and mammals have shown all imaginable combinations of relatedness between nonbreeding helpers and the youngsters they help raise. From this starting point, other discoveries have come to light through close attention to questions that had not been formulated prior to Hamilton's rule. For example, Belding's ground squirrels have extraordinary abilities for discriminating kin from non-kin, a possibility that had not occurred to anyone until it was clear how valuable such a skill might be. Warren Holmes, together with Sherman, showed that these squirrels can distinguish between varying degrees of kinship, such as full and half-siblings. Similar sensitivities have been demonstrated for many social insects, for barnacles, and, as we shall see later, for tadpoles.

Hamilton's rule thus opened our eyes to an evolutionary mechanism that promotes generosity and aid-giving, giving a much-needed lift to our understanding of natural selection. But his rule also specifies neatly where the border ought to lie between nepotism and selfishness. That is, when the rule's inequality ($br - c > 0$) is satisfied, altruism can spread; but when the inequality is not satisfied, organisms are expected to act on behalf of their own self-interest, as usual. In other words, the mix of relatedness, benefit to the recipient, and cost to the donor may predict manifestations of social behavior that are either nepotistic or precisely the opposite. When the rule is not satisfied, one might expect some rather nasty behavior even between very close kin. Think of this as the Dark Side of the Force. While Hamilton's rule gave us the sunny notions of cooperation and generosity arising through the indirect component of inclusive fitness, and the first generation of researchers understandably raced out to explore that as the chief priority, a second wave in the 1980s and 1990s helped provide the details. A full sibling is a glass half full, but it is also a glass half empty. We now recognize that in many circumstances individuals have to make tough choices between direct and indirect fitness, that is, between seeing to their own needs and catering to the needs of their close relatives.

Armed with this bit of background theory on the special properties of families, we can see that the diversity of biological phenomena should deliver some remarkable examples of sibling rivalry. Parental investment itself comes in widely different forms and is likely to be shaped at least in part by the effects it has on offspring selfishness. The interests of the parents certainly play a role, especially when they may be in disharmony with each other or at odds with the way their progeny are receiving investment. We shall have to look closely at who wins and who loses as these dramas of conflict are played out.

3

Nursery Life with Attitude

He can run, but he can't hide.

—Joe Louis

There are many species that never participate in families, but simply spew tiny offspring out into the world to fend for themselves. The ocean sunfish, for example, discharges up to 300 million fertilized eggs in a single oceanic spawning event, releasing them to drift off into planktonic anonymity.[1] That is an extreme, at least among vertebrates, but there are myriad alternative parental solutions, many of which involve creating fewer offspring, restraining them physically (either within a parent's body or in some external incarceration), and lavishing them with parental investment. The sand tiger shark, for example, produces eggs that not only remain inside their mother, but actually hatch there and swim about freely. More amazingly, the youngsters cannibalize one another *in utero*. The inch-long embryos develop an extensive array of touch sensors that help them feel their way around in the uterine darkness and find sibling eggs to swallow. Soon, their teeth start erupting and the bigger hatchlings can gobble up the smaller ones, a process that continues until only one remains alive. Even then the mother shark continues to shed additional eggs as food, an estimated 17,000 during the nine to twelve months of gestation. On this rich diet, the survivor grows to an impressive length of more than three feet and a weight of over twenty pounds at birth, an overall prenatal increase of 1.2 million percent.[2]

This gem of natural history was discovered quite unexpectedly, on July 27, 1947, by a biologist named Stewart Springer who was dissecting a ten-foot-long female shark. Springer reported: "Examination of the embryos began in a startling way. When I first put my hand through a slit in the oviduct I received the impression that I had been bitten.

What I had encountered was an exceedingly active embryo which dashed about open-mouthed inside the oviduct."[3]

The rich mixtures of cooperation and conflict that characterize family social relations are most likely to be manifest in the kinds of kin groups that remain physically together, as in a shark womb, rather than scattering immediately. A few years ago I set off on a library search for an appropriate label, an umbrella term that could be used for any organism, belonging to any kingdom, that stashed offspring in a spatially restricted site. I eventually selected *nursery,* the oldest of the candidates, apparently having retained this essential definition since at least A.D. 1300 for animals and since about A.D. 1565 for plants. "Nursery species," then, can stand collectively for all life forms that restrict offspring spatially at some early point in their developmental cycle. For *placental* mammals, such as dogs and cats, the first nursery is a womb, but the young of many species then spend additional time in a second nursery, the burrow or den. For *marsupial* mammals, like opossums and kangaroos, the second nursery is a pouch. For birds there are nests. For grebes, chicks leave the nest after hatching and ride around on their parents' backs, so those become nurseries.[4] For burying beetles, the nursery is actually the corpse of a rodent or a small bird, which the parents painstakingly denude of fur or feathers and modify into a malodorous ball that their larvae will eventually consume; this prize is interred an inch or two below the soil surface prior to egg laying. And, of course, not all nurseries have attendant nurses. A parasitoid wasp mother lays her eggs in or on the body of a paralyzed caterpillar, then departs. Upon hatching, her abandoned larvae share a finite resource base—the caterpillar's body tissues—and if that runs short, they must resolve the problem by competing among themselves.

With plants, the nature of the nursery varies widely, depending on what one counts as an offspring. For example, if the seed is considered the offspring (which makes sense because it holds the DNA necessary for building a future plant), it may be surrounded by a fleshy mass of fruit tissue either in isolation (plum or cherry) or with a few sibling seeds (apple or string bean). The fruit could be the nursery, but in the

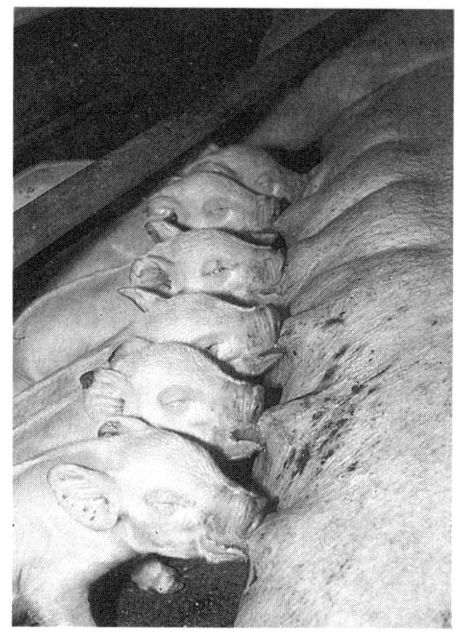

Any place where dependent offspring are confined can be considered a nursery. *Top:* Young western grebes ride on one parent's back. The other parent brings them food deliveries of small fish. (Photo: Gary Nuechterlein.) *Center:* A litter of piglets stays in a den (or a farrowing crate) where they can nurse from the sow. (Photo: David Fraser.) *Bottom:* To a seed, the apple itself is the nursery, providing protection and nutrients from the mother tree. (Photo: Douglas Mock.)

case of plums and cherries it may be more useful to consider each whole fruit as the offspring unit and count a cluster as the nursery (if that is the level at which the parental physiology for allocating nourishment operates). If water and nutrients ascending from the roots are distributed differently from photosynthate descending from the leaves, the competitive unit might even be a whole limb-full of sibling fruits, with perhaps the entire tree as the nursery.

It should also be evident that different kinds of nurseries persist for different amounts of time. A dandelion head seems the antithesis of a nursery when it is fully ripe, with its dozens of tiny white parachutists flying off independently and unlikely to germinate in close proximity to one another (though I doubt whether anyone has tried to follow marked individuals). But during the days when the seeds are developing and the dandelion flower is yellow, these offspring share a common resource base of water from the roots and energy from the leaves. Then it is a fine, if ephemeral, nursery. A red maple seed, with its whirligig wing that produces the familiar helicopter-like descent, usually travels much shorter distances than a dandelion gossamer. Logically, it should be somewhat more likely to germinate next to a bunch of maternal half-sibs. The resulting proximity means the potential for siblings to compete for resources may be extended to the full life span. For wingless seeds, like acorns and coconuts, that plummet more or less vertically, the odds of ending up among siblings should be even higher, unless squirrels or some other dispersing agents interfere. Gravity simply rearranges the three-dimensional tree-borne acorn nursery into a compressed two-dimensional form on the soil below. The identities of each seed's nearest neighbors undoubtedly are shuffled in the process, but most should come from the same tree. Once fallen, such clumped offspring no longer compete over access to maternal photosynthate, but they may well compete for soil nutrients and especially for growing space and sunlight with sibs and the mother tree alike.

The nursery for such plants is aptly called the "seed shadow." I once visited a wonderful field study of tropical forest trees at the Smithsonian's Tropical Research Institute on Barro Colorado Island, Panama, where nearly a quarter of a million individual trees have been tagged and censused for many years. The most impressive thing I learned that day from Robin Foster, one of the chief investigators, was that many of the spindly little saplings were actually fifty to sixty years old, but had landed in their mother's seed shadow and been unable to do much growing. Lacking sufficient sunlight, they entered a limbo lasting decades, as if waiting for their enormous mother to be struck down by lightning or compromised fatally by termites and carpenter ants. Sooner or later something like that will happen, and her venerable babies will then rouse themselves to engage in the arboreal equivalent of a furious vertical sprint as they grow into the vacated gap in the forest canopy. Even that sprint will take years to complete, and in the end there is likely to be room for only one winner.

Among warm-blooded vertebrates, which tend to lavish considerable parental care on their young, nursery tenure ranges from lifelong (many acorn woodpeckers never leave their California natal territory) to a brevity that would impress a thinking dandelion. The world's most precocial (or fastest-blooming) bird, the chickenlike Australian brush-turkey, is a fine example. The adult male rakes up a colossal pile of forest leaf litter, three feet or more in height and several yards around, that rots and ferments, generating natural heat in the process. The promise of this warmth is the male's courtship commodity, and various females visit only long enough to mate and deposit eggs on his heap before departing. The male sees that each egg is buried properly in the pile and then tends the clutch by playing thermostat, poking his bill into the mass and keeping the temperature at about 33 degrees Centigrade by judicious addition or removal of vegetation layers. When a baby hatches inside the mound, though, it is on its own. First it digs up and out, like a sea turtle, and then it starts preening. Whereas our familiar garden birds, such as robins, hatch naked or with just a thatch of insulative down, the young brush-turkey emerges with a complete feather coat, including full-sized flight feathers; after just two hours of drying, it can fly off and shift for itself. So, although the mound is clearly a nursery, with various siblings (paternal half-sibs) incubating inside at any given moment, sibling strife is unlikely because the siblings have no need to compete for anything prior to dispersal. Not all nurseries harbor conflict.

To summarize, when contemporaneous offspring are dependent on finite (and often parentally delivered) resources, there is at least the potential for some kind of shortfall. There is also a very real sense in which offspring compete with siblings that have not yet been created, if consuming limited local resources today detracts from what will be available in the future: today's credit-card debt can diminish tomorrow's solvency. Such non-contemporaneous sibling rivalry is commonly envisioned in terms of the wear and tear that expensive care imposes on parents.

Jumping from the world's most precocial bird to a comparable mammal, we start to find some potential for sequential sibling rivalry. A strong candidate for the title of fastest-blooming mammal is the wildebeest, a peculiar-looking ungulate of the African savanna that gives birth during the middle of a very active migration that covers many hundreds of miles. The pregnant mother is literally galloping across the plain one moment, surrounded by thousands of fellow travelers, then pulls off to one side and gives birth while the herd continues to hurtle past. The newborn struggles to its feet and wobbles briefly on pipestem legs, but it starts running within fifteen minutes, a handy feature considering that its mother (its source of milk) is ready to resume her trek. For wildebeests the nursery has no spatially fixed location and can be specified only as "near Mom," but this mobile bond must be maintained continuously because the nursery is moving at minivan speed across the Serengeti. And because litter size is always one, the only siblings with whom the calf might compete are either in its mother's past (if their demands diminished her ability to render proper care) or in her future (if its current selfishness might have that same effect).

So nurseries can be conventionally stationary, or mobile, or even a bit abstract in the sense of a present-future trade-off. The theme that cuts across all this variety is that parental resources are limited; when the shortage becomes acute, close kin may come into conflict.

As indicated earlier, natural selection is a reproductive race among individuals within the same population, based on and affecting traits that have at least a partially genetic basis. More precisely, it is a replication race among alleles within a population. Alleles that manage to copy themselves more rapidly than the alternative alleles at the same locus tend to be represented more strongly in the future composition of the gene pool. Allele-replication is largely a function of how individual bodies fare in breeding (the classical Darwinian view), but also can involve how individual bodies affect the reproductive success of close genetic relatives (Hamilton's extension).

Bodies accomplish gene replication in two general ways. Most conspicuously, they perform various behavioral acts, such as copulating and taking care of the resulting offspring. But nonbehavioral features also shape reproductive success, including things like how many offspring are produced in each batch, how frequently a new batch is created, the age at which breeding begins, the speed of growth, and so forth. Biologists call such characteristics "life-history traits" and realized some time ago that they, too, are forged by natural selection. Because plants do not have much, if any, true behavior (depending on how one classifies the rapid closing of Venus fly-traps and their ilk), virtually all of plants' adaptations for reproducing successfully lie in the realm of life-history traits. Thanks to time-lapse cinematic techniques, of course, we have all seen marvelous footage of vines racing up toward patches of sunlight and cacti blossoming "immediately" after desert rains, but that is an entertaining *trompe de l'oeil,* not behavior but growth.

It has become increasingly difficult (and less important) to keep a firm distinction between evolutionary perspectives on behavior as such (the field currently called "behavioral ecology") and on the nonbehavioral life-history traits ("evolutionary ecology"). For example, how can one coherently discuss the titanic battles among male elephant seals, as they bloody the beaches of Año Nuevo Island, California, without reflecting simultaneously on how selection has shaped their bodies for battle (fangs, callused chest skin, enormous bodies) and how it has shaped their mates differently? A male elephant seal weighs up to three tons, roughly triple the female size. In African lions, too, the males are much larger than females (about 400 pounds versus 250 or so). In both of these mammals, sexual size differences apparently evolved because success in combat provides great fitness rewards for males, while the payoffs for females come from other achievements (ability to capture prey, lactate, protect cubs, and so on). The optimal sizes for male and female bodies are simply different.

Because our real challenge is to understand the fundamental evolutionary processes that produce all kinds of traits, this blurring of traditional boundaries is healthy. Modern scientific journals and books merge the topics freely, and university courses that continue to be labeled "Animal Behavior" often contain considerable life-history material.

A key concept in the emerging hybrid field is that of the evolutionary trade-off. Any given phenotypic trait, behavioral or otherwise, typically has a mixture of positive effects on the lifetime reproductive success of its bearer (the benefits) and some negative ones (the costs). The net impact on fitness (benefits minus costs) is the bottom line that determines whether the trait tends to spread. The language used in framing these questions can cause semantic purists to shudder: we speak freely, for example, of the evolutionary "decision" concerning the age at which a male elephant seal should start fighting for supremacy on the mating beach. But the intention is simply to make clear that some array of options exists and that the underlying physiological mechanisms that govern when males enter combat, whatever those might be, will be judged as successful or unsuccessful by the final outcome, the lifetime breeding that is gained or lost. The shorthand of calling these "decisions" is merely a convenience for those of us interested in understanding more about the evolutionary or ultimate explanation for why such a trait exists.

A feel for the approach is easily gained through a simple thought experiment. Male elephant seals normally become sexually mature at the age of four years. We can imagine a hypothetical genetic change, an allelic mutation, that leads to acceleration by one year. That is, our mutant male avoids combat for his first two summers and enters the fray at age three. At face value, this might seem highly advantageous: such a male might commence his sex life a full twelve months before his rivals. He would also escape a full year of risk from predators like killer whales and great white sharks. By this I mean simply that a year of adolescence carries exposure to such dangers and is a time when regular (immature) males are not spreading genes. If that is converted into a breeding year, he is cashing in early. Surely such a genetic novelty might spread quickly. But we have addressed only the benefit side of the tradeoff so far. The down side is that our precocious mutant would have to face combat opponents that had had at least twelve extra months of body-building. The costs associated with picking a fight while still undersized could easily be fatal, leading to a lifetime reproductive success score of zero (and hence no gene-spreading).

I doubt seriously if anyone would try to take this project out onto the beach for a field experiment with real elephant seals, but the technique of surgically implanting tiny silastic tubes of time-release testosterone has been used frequently in recent years to "soup up" particular territorial songbirds. In general, these have shown that such males become more aggressive without necessarily being any better at winning the fights they enter. The thought experiment with elephant seal maturation can also be taken in the opposite direction. What if a mutant male's genotype directed its body to delay maturation until age five? Or it can be adapted for female seals (which, by the way, start breeding at age two and don't have to fight anybody).

Life-history traits shape the problems of supply and demand that arise in nurseries. Chief among these traits is the number of offspring created, the family's initial size. And there are several related issues, including how rapidly the young grow (hence how quickly they consume food and how soon they become independent), what sexes are produced, the interval separating consecutive breeding cycles, and so on. A beautiful example of how all this fits together has been shown for guppies in Trinidad, where David Reznick and his colleagues conducted a long-term field experiment to see how a population responds when its challenges are replaced with new ones. They transplanted 200 fish from the Aripo River, where they had evolved in response to a cichlid predator that targets adult guppies, to a small tributary that had lacked guppies but harbored a different predator, the killifish, which is known from other sites to specialize in juvenile guppies when they are available. In the ensuing 30–60 generations, requiring eleven years of repeated visits to check on the guppies, a strong evolutionary response was documented in the test population. Whereas guppies in the original river continued to mature early and produce large numbers of smallish babies per brood, as would be expected in light of the local cichlid's preference for adults, the group flourishing in the tributary came to defer sexual maturity and to produce fewer, larger offspring. One can easily imagine that behavioral changes may have accompanied these shifts (such as greater vigilance by juvenile guppies), but those were not measured.

4

The Trouble with Parents

Parents sometimes think of newborns as helpless creatures, but in fact parents' behavior is much more under the infant's control than the reverse. Does he come running when you cry?

—Sandra Scarr

In 1978, when I was a neophyte assistant professor writing my first grant proposal, a paper was published that cast avian family dynamics into a comprehensive evolutionary framework. Its title was "Brood Reduction in Birds: Selection for Infanticide, Fratricide, and Suicide?" Its author, Raymond O'Connor, pulled together theoretical perspectives from the work of W. D. Hamilton (inclusive fitness theory), Robert Trivers (parent-offspring conflict), Richard Alexander (parental manipulation), and David Lack (optimal clutch size). I embraced O'Connor's perspective with glee.

My objective in that first grant proposal, incidentally, was to explore how monogamous mating works, considering how easily natural selection might lead to destabilizing selfishness by two unrelated partners. As the most monogamous of vertebrates, birds seemed like good subjects. I chose to study the great egret because it is both monogamous and densely colonial (hence a potentially data-rich system). I also figured (incorrectly, as things turned out) that with such a large, white target it would be easy to dye-mark individual birds for identification.

O'Connor's paper called attention to the likelihood of serious sibling competition in birds. Virtually everyone in the field thought such competition was the province only of eagles and owls. But a photograph I had taken in an egret colony clearly showed the youngsters earnestly pecking each other in the face. Encouraged by this single scrap of evidence, I speculated that a solid, quantitative study of egret monogamy might also reveal that sibling rivalry occurred on a regular basis. Furthermore, noting O'Connor's assertion that "the fitness argument requires sibling aggression to be the outcome of conflicting sibling and adult interests," I figured there was an excellent chance that I would witness overt peace-enforcement by the parents. (Here I must credit my own parents for planting the seed of that possibility in their sporadic efforts to keep my three elder brothers from finishing me off during childhood.) Thus did I head off to South Texas in March of 1979, filled with great expectations, and towing a boat that possessed nowhere near the seaworthiness necessary for my chosen study site. I was ready in theory, but not in practice.

O'Connor designated three simple roles within a small family facing resource shortages and poised on the brink of an irreversible decision about trimming family size. His cast of players included Parent, Surviving Sibling, and Victim. Each party was seen to value the Victim's life from its own unique vantage, so they might agree or disagree on whether it should die.

What O'Connor predicted had been suggested informally by earlier authors, but he formulated the argument explicitly in mathematical expressions that specified just how severe conditions would have to be on the whole family for natural selection to favor sacrifice of the Victim. The first of these, which identified the balancing point or threshold for a nestmate to do the killing, showed that only modest levels of privation would be needed for the Surviving Sibling to give a thumbs-down. This O'Connor called the "fratricide threshold." Next, he showed that environmental conditions would have to be somewhat harsher for a parent to agree that it would be best for the Victim to die (interestingly, the parent's payoff is simply a dilution of the benefits gained by the Surviving Sib). Finally, he found a much higher mortality pressure above which even the Victim should favor its own death, the "suicide threshold." O'Connor wisely counseled readers not to hold their breaths waiting to observe voluntary suicides in nature, at least not with nestling birds, because the fratricide threshold will always be reached earlier and even the parents should turn on the Victim well before the latter is willing to throw itself on its metaphorical sword (that is, out of the nest).

If I were telling this story as a strict autobiographical account, the scene would now shift to egret nesting islands off the Texas coast. Each April for several years, I rented a house near Port Lavaca and hired field assistants from around the country to help me put in many hundreds of hours watching egret broods from plywood blinds. Basically, we maintained a vigil through the nestlings' first month of life, documenting all the meals and battles that occurred in our study nests.

But before telling those stories, I need to say a bit more about theory. I do this for two reasons. First, behavioral ecology is guided by ideas, the theoretical sketches of how we think nature might work, so it is appropriate to lay out some of that logic before relating how the birds actually behaved. Sometimes theory makes accurate predictions; other times it is completely wrong. Science matures either way. Second, it took several summers of observing, several winters of playing with the numbers, and several years of thinking about the puzzles before I understood much of what I saw. The actual fieldwork delivered many hours of wonder and excitement, but many more hours of tedium and trying to remain alert when the 90-degree heat and 90 percent humidity was promoting nothing more than grogginess. It would be very misleading to suggest that biological understanding comes quickly. Nature shows on TV do not portray much beyond the successful highlights from years of work.

In 1983 my wife, Trish Schwagmeyer, and I had the great good fortune of starting long-term collaboration with the theoretical biologist Geoffrey Parker at the University of Liverpool. We both wanted to learn about a set of mental tools called "evolutionary game theory," of which Geoff is a pioneer and master. In addition, Geoff had deep interests in both my work on sibling rivalry and Trish's studies of mating systems in ground squirrels. Trish and I figured we could haul our field data and intuitions over to Liverpool and learn about game theory by peering over Geoff's shoulder while he devised formal models based on our study systems and questions.

Game theory is a branch of what is grandly called optimality theory, which can be explained in terms of commonsense solutions to familiar engineering problems. In designing an airplane wing, for example, the angle of attack and the wing's general shape (area, length versus width, and so on) are chosen on the basis of the vehicle's intended use. If the plane is to carry heavy freight but need not be especially maneuverable, it will be given very different wings than if it, say, will be used to set speed records or land on tiny island runways. In all of these cases, the best (or optimal) decisions for wing design depend entirely on principles from physics and metallurgy. There is no opposing force with a vested interest in resisting the solution: gravity bears no grudges. Such exercises are thus commonly known as "simple optimality models" because there are no conniving rivals with which to contend (though the math is often not *my* idea of simple).

Game theory, unlike simple optimality modeling, explicitly takes into account possible counter-strategies employed by other living (and therefore evolving) players. For example, there is no such thing as the perfect first ten moves for a game of chess, for the obvious reason that your opponent is actively engaged in trying to defeat you: if you are predictable with a canned opening gambit, your rival adjusts to that reality and annihilates your position. Nature abounds with such games, since all forms of life are interacting in important ways with numerous other forms and all the living forces are subject to natural selection. For example, predators compete with prey over use of the latter's body tissues. Parasites are constantly co-evolving with their hosts, which are evolving defenses against them. Male members of a species are competing with rival males for access to females, and then their interactions with the females involve yet other kinds of games. In many such contexts, where two or more living players have conflicting interests, game theory is the logic tool of choice.

The term "game theory" reflects its historic origins in the topic of probabilities, which originally applied to parlor games. Consider straight-draw poker. Having received a new hand with four diamonds and one club, you assess the chance and impact of getting a fifth diamond on the draw (after discarding the club). There are at least two levels of puzzle here. The straightforward one is to calculate the odds of completing the flush, which requires only grade-school arithmetic. You can currently see four diamonds and you know that there are exactly thirteen in the deck, so it follows that there are nine possible winners out there that you might draw (the diamonds you cannot see). On the other hand, there are also thirty-eight other cards that would *not* fill the flush (the non-diamonds you cannot see). So your chance of success is 9 ÷ (38 + 9) or .19. Nearly one-fifth of the time you can expect to get what you want. But the other matter you must evaluate is whether your hoped-for success in the draw would prove to be a real success by winning the pot. That second problem is much trickier, as it involves your ability to assess whether the other players have stronger combinations of cards or will draw such before the final betting confrontation. Will a flush be adequate?

These two levels of decision make up the game. The first is clearly a simple optimality exercise, based on the inanimate deck of cards; the second is far more complicated because of the rival strategies involved. The other players may or may not have stronger hands. So the second phase involves what is called a "competitive optimality" problem and is also likely to involve various levels of deception during the discard and wagering periods that follow. One player may discard only two cards while holding a pair of kings, thus implying he has three of a kind. You are drawing just one card, but the other players don't know whether you're trying to fill a flush (and will have a worthless hand 81 percent of the time when you fail to get that last diamond) or have two pairs. Did he smile when he looked at his cards? Did you? Serious poker players monitor one another constantly, trying to detect what they call "tells," tiny unintended clues to key bits of information.

The probability tools took a short hop from the gaming tables over to the stock market. As with many forms of human endeavor, the lure of financial profit has attracted close study. Games associated with economic decisions have been modeled extensively. It should be evident that the success of my soft-drink stand will rise and fall as a function of how close to it you set up your competing soft-drink stand, how you price your wares, and so on. On a larger scale, an investor selecting a portfolio of stocks is playing against all the other investors, each trying to guess whether a given offering is under- or overvalued, trying to figure out when to get in or hurry out, and so on. The key characteristic in all these games, from the poker table to Wall Street, is that *the best move depends critically on what the other players are doing.*

This same principle applies to the behavior of animals, but there is an even more compelling reason for using game theory to study evolution. One great difficulty with applying game theory to economics lies in defining just what you are trying to achieve. It seems obvious that profit-maximizing is the goal, but on what time scale? You might set up your soft-drink stand and charge twice what I am charging, getting nice profits for a short time till the customers realize the rip-off and your business is shot. Conversely, you might undercut my pricing and suffer an indefinite period of no profit or even loss, in hopes of driving me out of business and making it up later on. The optimal strategy is not obvious because the goal itself is not clearly defined.

John Maynard Smith, a British geneticist originally trained as an engineer, recognized an extraordinary opportunity for adapting the game theoretical approach to biology. In the 1960s and 1970s he founded *evolutionary* game theory, which enjoyed one colossal advantage over economics: in evolution we really do know what is being maximized, namely fitness (that is, allele replication). Any trait or style of play—called a "strategy" in a nod to the field's parlor-games origins—that has the effect of enabling its body to out-reproduce rivals using a different strategy is favored by natural selection. The best of the array of candidate strategies tends to spread. The winning tactic from each imaginary tournament is said to be the "evolutionarily stable strategy" (ESS for short), which, by definition, is invulnerable to invasion by any alternative strategy until something comes along to disturb that equilibrium.

To get some understanding of how an ESS equilibrium is maintained, consider the sex ratio of some diploid sexual creature, say red-winged blackbirds. The red-wings' mating system is famously polygynous, with a few males mating a lot and many males not mating at all, while females are somewhat less variable sexually. Are red-wing parents better off raising sons than raising daughters? Specifically, we might ask how much parents ought to invest in sons relative to daughters. Intuitively, the best possible son:daughter investment ratio seems to be parity, for the fundamental reason that the species is diploid. Because every offspring has exactly one mother and one father, the overall average fitness achieved within the whole population of males must precisely match the average for females, unless there is a bias in the adult sex ratio to start with.

This may be clearer with a few round numbers. If the original numbers of males and females are roughly equal, say 100 of each in our imaginary marsh, and 300 fledglings are produced in a given season, then the average male success has to be 3.00, and so does the female average. The fact that some individual males may have scores of 30–50 while lots of others have zeroes does not affect the male average. Imagine now that the Angel of Death visits the marsh, perhaps a Cooper's hawk that tends to kill more of the exposed and conspicuously colored males than of the relatively furtive and drab females. If half our adult males die suddenly, we are down to 50 male blackbirds mating with the 100 females. Suddenly, the male average success doubles to 6.00 and a substantial fitness incentive materializes for parents to invest preferentially in sons: under these new conditions, raising sons will be a more profitable strategy, generating twice as many grandkids as raising daughters. Under these new conditions, anything that promotes enhanced investment in sons should be favored. Selection is said to be "frequency-dependent," meaning that the payoff for one strategy (here, skewing investment toward male offspring) hinges on how common it is in the system. So birds with a predilection for producing sons will enjoy a temporary pay raise and their genes promoting that bias will tend to spread, at least initially.

The beauty of this logic is its power for self-correction. In our example, the bias toward investing in sons automatically pares away the extra payoff accruing from that very activity. Imagine that after a few breeding cycles of favoritism toward sons, the adult population has moved to 100 females and 75 males, a net gain of 25 males from the low point. Now there is still a modestly higher profit to be had from building sons over daughters, but it is no longer anywhere near the twofold advantage. The mean payoff available from a son has fallen to 4.00 grandkids. As sex ratio parity is approached, the payoff asymmetry diminishes and eventually vanishes altogether. And once that equilibrium has been restored, the ESS is once again returned to equal total investment in male and female progeny.

Back in Liverpool, Geoff Parker began applying game theory methods to the problem of fatal sibling rivalry in egrets. We focused initially on what the system would be like if the key limiting resource (parentally delivered food) were distributed among nestmates according to the best interests of the socially dominant elder siblings. That is, if the senior sibs were able to exert despotic control over how much they could eat, what would be their preference? It was already clear from my field time in Texas that the bullying elders seemed to be getting their way, so this was a reasonable starting point.

In a brood of great egrets there are typically three chicks that hatched at one- or two-day intervals, such that the first-hatched (the A-chick) is older and stronger than the second (B), which has the same advantage over the youngest (C). If A enjoyed hegemonic control over the food, eating all it wanted before leaving the remains to its siblings, then B could take as much as it liked, and so on, how would the food be divided? Geoff's model suggested that A's selfish fitness interests should impel it to take more than its fair share, but that its piggishness would eventually be constrained by its kinship with its nestmates. If A had only one sibling, the optimal solution would be for A to consume all the available food until the point where the next mouthful would do A's personal fitness just half as much good as it would do B's (because A has a 50 percent stake in B's fitness through shared genes). But that is not a "game," really, because so far only one player gets to do anything; adding a third nestmate to the model creates an extra layer to the problem. Now A continues gulping down the food until the point where its personal fitness gains almost nothing and so it does not grab the next item, which B promptly eats. This patterns persists for a while as B's situation improves, until a point is reached eventually where A's fitness would be best served by having C eat something. After all, C is as closely related to A as B is. But that's not how the world is ordered. The extent of A's control is limited to the digital decision of whether to eat the next item or not, whereupon the choice is B's alone. Thus, if A does not eat the next item, it will not go to C, but to B. It can be shown easily that in this situation A's fitness will be better served by resuming its own consumption than by letting B have the food. And we can see that this dynamic between A and B, this game within a game, is very hard on C. Specifically, if our guesswork about which parties are in control is accurate, then we expect three-chick egret broods to show a particular pattern of food distribution, with A getting the most and B a close second, but poor C quite a distance behind these two.

This may seem a bit like ancient philosophers debating how many angels can dance on the head of a pin, but it isn't. Two other old arguments offer better parallels, namely the ones concerning how many teeth a horse has (supposedly disputed for two weeks in 1432 until someone suggested checking a horse) and whether tomatoes (a.k.a. death apples) are poisonous. It is possible, after all, to count equine teeth and eat a tomato, a pair of empirical exercises that eventually settled these controversies. For the egrets, my assistants and I had accumulated detailed records from many broods, where we had estimated the size of every regurgitated bolus of fish and noted who ingested it. After the model was completed we tallied these data, which showed that A-chicks had averaged just over 43 percent of the total food, modestly ahead of the Bs' 37 percent share, but the two of them far ahead of C, which got slightly less than 20 percent.[1] This suggests (but does not prove) that much of the food distribution in egret broods is indeed controlled by the offspring.

The interplay between parents and their progeny is extremely interesting because of two built-in asymmetries, one physical and one conceptual. The latter was first identified by Robert Trivers in a classic 1974 paper entitled "Parent-Offspring Conflict." Before Trivers, most authors addressing interactions between parents and offspring had considered only the positive, nurturing side associated with care given to the young (Nature with a Happy Face). Trivers's insight, clearly derived from Hamilton's rule, exposed the negative aspect. In brief, he argued that a parent should regard all its youngsters as being of equal value because its coefficient of relatedness with each of them is always .50. All else being equal, then, a parent should achieve its optimal allocation of parental investment by doling out essentially equal portions to all its progeny. From the point of view of an individual offspring, as we have seen, the coefficient of relatedness to a full sib is only half as high (.50) as relatedness to Self (1.00), so each youngster should prefer that the parents skew their investment in its favor. Thus there should be an automatic mismatch of preferences, with natural selection generally promoting greater selfishness by the offspring than the parent, looking out for its own best interests, should tolerate.

The early reaction to this argument provides an interesting illustration of why purely verbal reasoning is often inadequate. As soon as Trivers's paper appeared, it was rebutted by Dick Alexander, who argued that any genetic novelty leading an offspring to be harmfully selfish (to a degree that reduced its parents' fitness) could never spread. The explanation was simple in words: such a selfish offspring would, in time, become a parent and then the trait's effects would backfire; its own offspring would use the ultra-selfish trait they had inherited to make *it* work extra hard to serve *them.* As a graduate student in 1974, I remember thinking Alexander's position made sense. But I had also thought Trivers's logic made sense. And in 1976 Richard Dawkins offered yet another way of describing the problem in his immensely popular and influential book *The Selfish Gene.* He thought it obvious that Trivers was right and Alexander was wrong. Annoyingly, his argument, too, sounded right to me. With this swirl of incompatible explanations confusing the issue, I was relieved when theoreticians leapt in to set things straight.[2]

Let me offer what I think is an easy mathematical explanation for understanding whether parent-offspring conflict can evolve at all.[3] Imagine the transformation of an allele we can call *g* into a dominant mutation, *G,* that has the effect of leading the offspring body in which it resides to develop an extra increment of selfishness. As a result, the selfish offspring's body takes more than its "fair share" of parental investment, raising its fitness by some amount. That is, a heterozygous *Gg* individual gains some tangible benefit that must be subtracted from the finite pool of parental investment. Once consumed, that investment is unavailable to other offspring. To keep things simple, imagine that a mated pair produces just two youngsters, which are created sequentially, as their entire lifetime output. Thus we have two categories of offspring, a typical type (*gg*) and a selfish type (*Gg*), plus two temporal roles (a given offspring can be either first in the family order or second).

Within this framework, there are just four possible kinds of families:

one in which both offspring are typical *gg*, a second in which both are selfish *Gg*, and two others that contain mixed genotypes. Obviously, birth order does not affect things in the first two types because the kids are identical with respect to our focal gene. In the two mixed sibships, either the older or the younger is selfish (so we can have either *Gg* followed by *gg* or *gg* followed by *Gg*). Now look at the effects on offspring fitness. In the all-*gg* family, each offspring receives the standard amount of parental investment and realizes the standard fitness return. We can symbolize each offspring's fitness benefit as w and summarize the parents' total benefit as being $2w$ for the two kids. In the mixed family where the typical sib comes first, things are exactly the same: the second offspring has the allele for extracting extra investment, but it lacks the opportunity to do so because of its birth order (since, according to our assumption, the older sibling gets first choice of the food the parents provide).

In the two remaining family types, however, selfishness takes its toll. The first selfish offspring in the all-*Gg* family extracts its bonus (of value b) and enjoys an enhanced fitness ($w + b$), while the second selfish sib suffers whatever cost (of magnitude c) is associated with the diminished residual investment, so its fitness is lower ($w - c$). Finally, in the *Gg*-then-*gg* family, the later-born sib is not a carrier of the mutation. In both of these family types, parental fitness is $2w + b - c$, so we can see that if the costs suffered by one offspring are larger than the benefits gained by the other (if c is greater than b), parental fitness is reduced.

The remaining question is whether the selfish *G* comes out ahead of the typical *g* in terms of allelic replication rate. The above results reveal that it must do so, on average, because only *G*-bearing offspring ever acquire the fitness bonus b, while both the *Gg* and *gg* genotypes share in paying the costs c, depending on whether the first sib in a family is selfish. I can see now why Alexander's verbal argument bamboozled me: it contained the unstated assumption that c must equal or exceed b. If that is so, then he is correct: parental fitness cannot benefit from offspring selfishness because $2w + b - c$ will always be lower than $2w$. This general conclusion also shows that there may be circumstances under which parental favoritism toward certain offspring (to the detriment of others) should evolve.

The model just described, fashioned by Judy Stamps and Robert Metcalf of the University of California, plus similar models by Geoff Parker and others, effectively resolved the initial controversy surrounding parent-offspring conflict theory. Selection really does seem capable of producing offspring selfishness, even at levels harmful to parental interests. So now the matter of interest shifts to which side, parent or offspring, prevails when a significant mismatch of interests arises. For that matter, showing that a theory is logically sound (that the process it proposes can exist in principle) is all well and good, but the really important question is whether the theory actually clarifies anything that would be murky otherwise. In this case, the key issue is whether theories of parent-offspring conflict shed helpful light on the real interactions of parents and their progeny. We will revisit that issue in Chapters 11 and 12.

A whole constellation of resource-shortage problems would cease to exist if parents simply doled out enough investment to meet the needs and desires of all their progeny: selfishness beyond that point would accrue no advantages and thus would become moot. At the extreme, parents might produce a single offspring and spend the rest of their lives making it comfortable and successful. But alleles predisposing parents to breed so slowly would be replaced in short order by alternative alleles for higher fecundity and more prudent, limited investment in each youngster. A one-child-per-lifetime policy could never evolve under natural selection. Following this logic further, we can quickly see that parents commonly need to err on the side of producing somewhat "too many" offspring—more than they are likely to be able to provide for—because alleles promoting that strategy are more likely to maximize lifetime reproductive success, even after all costs associated with sibling competition have been paid.

Across the plant and animal kingdoms, parents gain three general fitness dividends from overproducing initially.[4] These are sometimes mutually compatible, allowing parents to collect from more than one source. First, by creating more offspring than they can normally expect to raise (a portion of the family hereafter called "marginal" offspring), parents can capitalize on abnormally favorable conditions for raising young. The idea here is simply that parents must commit to a particular number of kids in their opening move, which they must make days, weeks, or months before the unpredictable ecological realities unfold. If parents aim high at the outset and the season turns out to be exceptionally rich (say prey is plentiful and easy to find), they can take advantage of that good fortune. Conversely, if the season turns sour, so not all the offspring can be raised properly, various steps may be needed to reduce family size. Thus initial overproduction, when combined with some mechanism for subsequent brood reduction, offers parents a two-step method for achieving the highest possible breeding success in an uncertain world. The gambit effectively allows them to track the resource base.[5] An alternative approach to fluctuating ecological conditions would be for parents to abandon the current family and start over again when they realize they have produced too few offspring. This has the obvious disadvantages of being slow and wasteful.

Consider a pair of egrets normally capable of raising, say, two healthy chicks in a typical season. If they produce just two eggs and it turns out to be a rich food season after the first two have already hatched, the option of adding third and fourth eggs will involve restarting the clock. Even if they keep the two they have, time is still needed to metabolize the new eggs (about a week) and incubate them (3½ more weeks) just to reach the same point in the cycle. Furthermore, they will now have a two-tiered family, the first pair of chicks being a month older than the second pair. And by then, of course, weather and food conditions may have deteriorated. A bit of early parental optimism simply guarantees that family size will not err unduly on the side of conservatism, though it may also extract some costs associated with eliminating one or more marginal offspring later. So the first class of incentive for parents to overproduce, which we'll call "resource tracking," is all about getting the *quantity* of offspring right.

A second problem that parents can sometimes solve by overproduction relates directly to offspring *quality:* any offspring may turn out to be developmentally defective in some way that reduces its value as a generator of grandchildren. One extreme in the continuum of imperfections is death. If a core brood member dies early on, for whatever reason, the presence of a backup marginal offspring provides an insurance policy. Short of dying, a core offspring may just happen to be wimpy and undeserving, because of either intrinsic flaws (perhaps bad match-ups of parental genes such as deleterious homozygous recessives) or acquired flaws (exposure to disease, accidental injury, and so on)—anything that converts a first-string youngster into a bad bet for full-term investment. In such circumstances, a marginal sibling may be able to switch roles with the wimp. This happens on sports teams all the time, and it is why fourth- and fifth-string players are kept on the roster. If a starter is harmed, for example, by an "accidental" elbow or eye-gouge, a scrub is available. In an egret family, there are occasional hatching failures and grab-and-run predator raids that enable a marginal offspring to become a core family member through no action of its own. That is, a last-hatching chick may end up a star in the family without having to demonstrate any special skills, so long as it sticks around. The second class of parental incentives for overproduction, therefore, pertains to the value of marginal young as "replacement offspring."

As an extension of the insurance idea, parents could make the decision to substitute a marginal brood member for a core sibling if they performed some kind of assessment process and deemed the scrub to be worthier than the original starter. If so, the substitution would upgrade the average quality of the brood and enhance parental fitness. Treatment of the displaced original core sibling might be harsh (execution by either the parent or newly elevated nestmate) or rather subtle (redirection of parental favoritism). It seems quite clear that many plants employ selective forms of spontaneous abortion so as to reap this benefit (see Chapter 13). Whether animal parents also do this is less certain, but provocative evidence is beginning to suggest that some may.[6]

The third general class of parental incentives, called "kin facilitation," addresses the fact that extra offspring may provide some valuable services to parental fitness other than (just) making grandkids. In various social animals, including acorn woodpeckers, non- or pre-breeding offspring serve as helpers, providing real assistance to their parents in the rearing of younger siblings. Such nepotism reaches extremes in certain highly specialized insects, like honeybees, where most daughters are permanently sterile and contribute all kinds of labor to the hive's economy. Their efforts liberate their mother, the queen, from maintenance chores, and she transforms her body into an immobile egg factory, producing mainly additional cohorts of sterile workers. Periodically, small batches of new larvae are raised a bit differently, in larger cells and with a richer diet, such that their gonads develop fully. In time these can become breeders.

In less social species, additional marginal offspring in the nursery can also serve more limited functions. In the early spring, for example, when nights are chilly, an extra chick in a brood of birds can make the thermal micro-environment more favorable by helping to insulate core-brood siblings, reducing their surface exposure and saving them energy. This is not altruistic, since its own temperature maintenance costs are similarly lowered, but from the parental perspective it may help tip the balance toward adding that one additional egg to the nest.

At a higher level of sacrifice, marginal offspring are sometimes consumed when all other food supplies are exhausted. The possibility that parents may be counting on sibling cannibalism is often referred to as the "icebox hypothesis." Nobody seems sure where the term originated, but some have attributed it to Dick Alexander, who wrote to me that, in his boyhood on an Illinois farm, it had occurred to him many times that a chicken, being just about the right size for a single family meal, was the functional equivalent of meat in the icebox. So, at least in principle, in overproducing offspring parents could be converting early-season protein into a stable form that will retain its freshness for a subsequent famine.

It is easy to see that these three classes of parental incentives can sometimes operate concurrently. For example, if the season does not turn out to be unusually rich, such that the once-viable resource-tracking option has collapsed, a marginal American kestrel chick still represents insurance until the moment it is (irrevocably) dead, at which point it is often eaten.[7]

The nursery takes many forms and receives broods of many different sizes across animal and plant taxa. Whereas we are accustomed to thinking of nurseries as protective cocoons, they also have sinister po-

Barn owls produce unusually large families in which nestmates are often one to three weeks apart in age and thus considerably different in size. If food turns out to be scarce, larger chicks have a competitive advantage over their junior siblings, but they are not known to attack them physically. (Photo: Carl Marti.)

tential that is related to the spatial confinement of their occupants. Like the ropes that delimit a boxing ring, the nursery has spatial boundaries. And, while true sibling competition may or may not arise during a given breeding cycle, we have seen that parents routinely err on the side of overcrowding the system. The unpredictable nature of resources complicates life in the nursery. If resources are sufficiently likely to become scarce at some point in the near to middle future (any time before the young disperse), natural selection can favor a biological preemptive strike that foists the misfortune onto the least valuable family members.

5

Raising Cain

That all men should be brothers is the dream of people who have no brothers.

—Charles Chincholles

One emerges from its shell into a shallow scrape in cold Canadian prairie soil, the other hatches on a warm cliff ledge in Zimbabwe. They have just escaped the rigid shell prisons that simultaneously protected them and limited their growth, but both are already in really, really serious trouble. The former is a hatchling American white pelican; the latter, an African black eagle. In both cases the source of their trouble is a deceptively warm and fuzzy nestmate that hatched a few days ago and has put the interim to good use, building size, strength, and coordination that will soon be used to kill the new hatchling. Blinking and looking around for the first time, these new junior sibs have only a slim chance of surviving the next week.

In the eagle nest, the weapon of execution is a strongly hooked beak, the device that will be used throughout the rest of the killer's life to dismember the bodies of prey it has seized and killed with its powerful claws. Black eagles are particularly fond of rock hyraxes, an abundant jackrabbit-sized herbivorous mammal that lives colonially in the granitic outcroppings (kopjes) near the eagle's nesting area. But the youngster's feet are not yet dangerous, so it must use its bill to peck and tear at its sibling, opening flesh wounds and inflicting pain, quickly intimidating the subordinate to the point that it stops even begging for food. Death follows within a day or two.

The whole process of black eagle siblicide seems to have been observed in its entirety only twice, first by E. G. Rowe over fifty years ago and later by Valerie Gargett, unquestionably the doyenne of black eagle biology.[1] This seeming dearth of documentation is due to the ordeal involved in making such observations. You must build a platform in a nearby tree (or erect a tower that the birds learn to accept) with a small blind, then endure at least several consecutive long days totally inside the blind (to avoid spooking the parents), watching and taking notes. Assuming that you time everything perfectly and enter the blind before daybreak on the day the B-chick hatches, to get the whole story you still have to be there for all daylight hours until the victim has succumbed. And after all that effort your sample size is just one. There is a paucity of observers with sufficient patience to complete this exercise, so the literature's sample size stands at two.

Mercifully, the two accounts are straightforward and much in accord: in each case the A-chick simply went on the attack within hours of B's hatching, especially when B tried to receive a scrap of meat proffered by the attending parent, and continued sporadically until B was dead. One of the most perplexing aspects of the attacks is that food did not appear to be in short supply at all: Gargett meticulously reported that a full 5.7 kilograms of hyrax carcasses were littered around the nest she watched. By the time that particular B-chick died, three days after hatching, it had been assaulted on 38 separate occasions and Gargett had counted 1,569 distinct blows of A's bill. During this time, A had packed on another 50 grams of body mass, while B had withered away to die weighing 18 grams less than what it had weighed at hatching. After B's demise, six hyrax carcasses could still be seen at the nest.

We have rather fewer gruesome details for pelican siblicide (surprisingly, since pelicans nest in dense colonies and should be easier to observe than black eagles), but their process seems only slightly less harrowing. Once again, the weapon is the A-chick's bill, but this one is long and straight. The pecks are also straight, lunging jabs hard enough to create visible wounds that quickly intimidate B. At high latitude, chicks require parental brooding against the cold, and A's bullying enables it to take the preferred front position by the parent's breast. While being brooded, with the tending parent settled lightly but firmly atop them both, A cannot harm B, but the attacks are renewed almost every (94 percent) time the parent stands up. To feed its young, the parent pelican regurgitates partially digested fish up into its throat sac, then opens and extends its bill, nine times out of ten toward the chick at its breast.

A young golden eagle stands over the corpse of the sibling it has just pecked and starved to death. (Photo: Michael Collopy.)

Death seems to come more from starvation than from physical beatings directly, but that is necessarily an impression: the relative contributions of deprivation and injury are hard to determine objectively from an observation blind.[2]

In both of these species at least one parent is often present in the nest during the fighting, but the parents do not intervene to stop or reduce the carnage. Nor do they show special interest in getting food to the underfed and dying victim. This is one of the great enigmas of the whole phenomenon: in species after siblicidal species, parents tend to stand by passively while one vehicle of their Darwinian fitness destroys another. Initially, this made no sense whatever to observers. As I admitted earlier, theoretical considerations had primed me to expect dramatic interventions by the adults, but the birds were ignoring perfectly good theory!

The unflinching type of sib-killing practiced by white pelicans, black eagles, and around two dozen other avian species—wherein two eggs are laid but one chick almost always slaughters the other—is called "obligate siblicide." It flummoxed naturalists for various reasons. A brief historical sketch of this corner of the field may be helpful for understanding why the most obvious explanation, that the second egg serves as a backup or insurance against loss of the first offspring (originally proposed by E. F. Dorward in 1962), has been resisted strenuously by some authors.

Denial of the insurance argument seems to have stemmed from perhaps four misunderstandings. First, siblicide was initially assumed to be merely a necessary step toward cannibalism, as indeed it certainly seems to be in some non-avian species (such as the sand tiger shark embryos encountered in Chapter 3). The appeal of that reasoning may stem from its featuring an identifiable and tangible reward gained, a meal. This assumption seems to have created a strong mind-set, a logic trap really, and led to some interesting speculations that were unhampered by much evidence one way or the other. For example, Collingwood Ingram wrote about short-eared owls' habit of keeping what he called a "larder" of prey carcasses near the nest.[3] His accounts, based on visits to just two nests, described a small collection of rodent prey stacked neatly in repositories just inches from the owls' ground nests. Ingram creatively interpreted the larders as representing a pressure-release system for deferring sibling cannibalism, which he naturally assumed was "bad": the senior sibs, being the only ones developmentally capable of climbing over to the larder, would presumably help themselves whenever fresh prey deliveries were disappointingly slow, thereby assuaging their hunger and sparing their sibs. Ingram's interesting and plausible speculation may even be true: nobody has done detailed studies of short-eared owl chick behavior to this day, four decades later.

The second fallacy was a fundamental misunderstanding of how natural selection works, specifically the units on which its scouring action is most potent. Before the mid-1960s, selection was widely believed to operate for the good of whole species or communities, a concept called "group selection." As a result, Nature came to be viewed vaguely as rather easygoing and benign. True, there were some exciting and somewhat disturbing film images of cheetahs dashing across the African savanna to tackle gazelles, creating great explosions of dust and fur. TV viewers were beginning to be titillated by scenes of "Nature red in tooth and claw" on shows like *Wild Kingdom,* but the nastier struggles of life in the wild were nearly always depicted as operating *between species.* Indeed the narrators of such TV footage sometimes offered a palliative for what they may have imagined to be audience sensitivity, suggesting that the gazelle's death was all part of Nature's Grand Scheme and panning quickly from the thrilling kill to a scene of happy, healthy, and clearly deserving cheetah cubs whose welfare was comfortingly secure because of their parents' exemplary fleetness of foot. The gazelle's death was not in vain, the audience was reassured, but merely a necessary and brief moment of unpleasantness toward the support of cheetahs.

By contrast, fatal unpleasantness *within species* was seldom mentioned and unpleasantness *within families* was anathema. If selection operates mainly on whole populations, species, and the like, then infanticide and siblicide must be explained as some sort of editing process that removes the less favorable traits from the gene pool, thereby preserving the better alternatives, just as weeds are removed from the garden to enhance the peonies. The problem with this logic is that it requires a prescient gardener, or at least a force that is "building toward" some goal like a weed-free peony garden. If left to natural forces only (and remember that our topic is natural, not artificial, selection), your average weed kicks peony butt.

Back in the 1960s, some cases of within-species cannibalism had been described for a few animals, but its diversity was not yet appreciated, a lack of information that made the few hard-to-explain cases easy to sweep under the rug.[4] For a while the reports of hawk and owl nestlings killing sibs without eating them trickled in and were summarily dismissed as pathological or at least atypical. But the time-honored solution of simply ignoring facts only works while those inconvenient facts remain few in number. The mystery deepened with the gradual accumulation of anecdotes. How was the essential harmony of nature served by siblicide? It seemed highly unlikely that a practice in which roughly half of the chicks killed the other half would enhance the whole population: indeed, in rare and endangered species, obligate siblicide must have looked like a nightmarish express train to extinction. Whooping cranes, for example, got down to a few dozen individual adults, but their first-hatched chicks continued to eradicate B-chicks (a feature that federal research biologists presumably used to justify confiscating B-eggs for artificial rearing in the recovery program).

One way out of that conundrum was to focus on the possibility of competitive release, the argument that killing a nestmate might be essential, thus potentially beneficial, because of food shortages. Enduring confusion ensued because this argument seemed at odds with the few facts available: piles of hyrax carcasses were hard to reconcile with the notion of life-threatening shortages. Logically, the combat could hardly be "for" limited food when it was conducted amid great piles of food. The subtler possibility of a preemptive strike, which presupposes a pending competition over a future risk of food shortage,[5] struck some as a desperate and dubious explanation. Even in white pelicans, for which no conspicuous piles of food have been reported, common sense suggests that two ponderous adults, whose pouches are many times larger than their hatchlings, surely do not break a sweat keeping up with the needs of two tiny chicks during the early (pre-siblicide) days, especially when the lone victor consumes much greater volumes of food when just a bit older. Of course, this is the whole point. Chicks do grow very rapidly and their appetites enlarge accordingly.

At the root of the third fallacy was the failure to distinguish between proximate mechanisms and ultimate functions. The term "proximate mechanism" in this context refers to the immediate constellation of circumstances and physiology that leads to a particular behavioral act. This includes such things as the stimuli that trigger the response, the actor's sensory nervous system that receives and processes the incoming stimuli, the actor's central nervous system that makes some "decision" about how to respond, and the actor's musculoskeletal system that produces the behavior. In short, the proximate issue boils down to *how* a behavior occurs and mostly covers a rather short time frame, say ranging from microseconds (neural activities) to months (hormonal changes like puberty). By contrast, the "ultimate function" of the same behavioral act concerns *why* those proximate mechanisms evolved in the first place. We may marvel that world-class athletes can run at amazing speeds or leap to astonishing heights, but with a bit of reflection we realize that the behavior patterns themselves, running and jumping, were shaped for nearly our entire lineage's history by the need to chase prey and elude predators (including other humans). It is important to keep clear on the proximate-ultimate distinction when discussing behavior (and other traits as well), because both levels of analysis are important. For example, if you ask why cheetahs are so fast, you are presumably asking for an ecological key (because they prey on fleet gazelles), but the question could be answered on the alternative, and rather less interesting in this case, proximate level (because they move their legs very rapidly, have fast-twitch muscles, carry little body weight, and so on). The latter is really a "how" question (How do they manage to go so fast?), disguised with a "why."

Getting back to early siblicide, then, the net fitness benefits gained by launching a preemptive attack and getting rid of a future food-rival (and doing so while the cost of killing it is still low) can be disengaged completely from the proximate situation (piles of hyraxes). In short, natural selection operates on eventual outcomes, not present appearances. It is the supreme bottom-line business. If, on average, a food shortage will arise some weeks down the developmental road, or even if that happens only once in a very great while, selection may still favor a rapid execution. Anyone who has ever been forced to purchase flood coverage for a house in a 100-year flood zone will have worked through the reasoning behind this, however unhappily in the short term.

Finally, the insurance-egg argument was (and, in some circles, still is) resisted for many years because a widely cited data set seemed to refute it directly, using real numbers. We field biologists pride ourselves on being down-to-earth and sensible empiricists, with honest mud and excrement on our boots. Accordingly, data that we can look straight in the eye tend to be more persuasive to us than mere air-castles of logic. These particular data purported to show that B-chicks survive far too infrequently for natural selection to have favored the parental overproduction of a second egg. This important evidence featured our friend the black eagle and bears a bit of scrutiny because it contains the fourth fallacy.

At the heart of the problem lies the matter of how one measures the value of insurance. In our own decisions about life, health, and property insurance, this seems straightforward enough: we look at the cost of a given policy (the premium), the likelihood of calamity, and the value of the promised payoff from the insurance company (cynics will add another ingredient: the chance that the company will find some excuse for refusing to honor your claim). Similarly, for avian parents to create a second egg as backup, the total costs for its construction (ideally, measured in terms of lost future reproduction) must be outweighed by the eventual success *of that same egg* (how often its chick fledges, joins the breeding population, and so on). To take one extreme, if both chicks *always* hatched and then A-chicks *always* killed B-chicks, insurance could not possibly be cost-effective: that B-egg has no value if the first-stringer is a guaranteed success. But, of course, real first eggs do sometimes fail to hatch and a proportion of A-chicks can die very early from some other causes ("very early" meaning before B is killed), in which case the policy might be potentially valuable. Besides, the single premium might be cheap. It is very hard to say just how much an egg that weighs between 2 and 2½ percent of female body mass costs the mother to build, and it is equally hard to specify a threshold frequency for B-chick survival that would compensate for such costs, but the problem is tractable.

Toward that goal, Leslie Brown and his colleagues located 120 records of two-egg black eagle nests, ruled that only 2½ percent of the eventual survivors had begun their lives inside B-eggs, opined that such a low rate of success must surely be insufficient, and pronounced that the creation/destruction of black eagle B-chicks was "an inexplicable example of apparent biological waste."[6] I do love the candor of that passage and I credit the authors with having the guts to say their piece without mincing words, but there is a major flaw in their calculation that just might make the whole thing explicable. As mentioned earlier, there are good reasons why biologists do not spend lots of time hanging out near black eagle nests, or even visiting them regularly. Most of these particular records had accumulated over the years from occasional visits by amateur and professional eagle buffs, and their sketchy nature leaves problems in exactly how to interpret any perceived trends in the data. For example, these notes documented both chicks as definitely having hatched in only 44 of 120 clutches. More important, the persons visiting the nests apparently did not mark any of the eggs individually (a routine procedure for avian field studies nowadays, but one that necessarily takes at least two visits per nest to be of any value). Whatever the reason, there is simply no way of knowing which chicks came from which eggs. What to do? The authors solved this problem by "assuming that it was the second-hatched chick that died" in all nests that lost one of their two chicks.[7] Of course, such an assumption obliterates the potential for a B-chick to be given credit as a successful replacement.

Going back over the data and making what I hope are neutral assumptions, I find that approximately one B-egg in five lived to fledge, a 20 percent ballpark figure that fits well with other obligate siblicide species.[8] Using a smaller subset of her total nests, Gargett cited 5 B-chick survivors in a sample of 22 nests, which works out as 22.7 percent.[9] Of course, we still don't know whether such an increase in the estimate of insurance value would end up making the black eagle B-egg cost-effective as insurance, but it seems much more likely. Sadly, the 2½ percent figure of insurance payoff for black eagles continues to be cited in the literature.

It occurred to me that the discovery (in the late 1970s) of siblicidal behavior in several colony-nesting birds offered excellent opportunities for testing the insurance hypothesis experimentally, and I tucked that suggestion into my first review paper on siblicide.[10] What was needed was a simple removal experiment. Taking a colonially nesting obligate-siblicide species (I proposed choosing white pelicans or masked boobies), the question seemed obvious: would nests from which the B-egg had been heisted artificially be less likely to produce their single fledgling than nests allowed to retain the putative backup? If so, the difference would reveal how often B is allowed to step up and fulfill its potential. If not, the hypothesis would be in serious trouble. Roger Evans of the University of Manitoba quickly wrote to announce that he and his student, Kevin Cash, had figured this out independently, had performed the indicated experiment (on white pelicans), and would be happy to let me see the answer.

Actually Cash and Evans turned out to have done *both* of the appropriate studies. In one sample of nests they had not removed any eggs, but had carefully marked each egg on the day it was laid and later applied individualized dye marks to the hatchlings' downy coats. These steps eliminated the need to make dubious guesses as to who was A, who was B, and which one died. In a second set of nests, they experimentally manipulated egg numbers by moving second eggs from one group (to make one-egg clutches containing A only) and slipping these into the nests of a third group (creating three-egg clutches of resident A and B plus an alien B′), while leaving a third comparison group (the controls) with their normal clutches of two. In these modest samples, 18–20 percent of B-chicks fledged.[11]

The difference between how the black eagle and white pelican systems have been explored provides an instructive example of the way scientific progress often runs in field biology. A colony crammed with pelican nests every few yards offers a very data-rich environment for scientists. A field biologist can sit quietly inside a blind and keep watch over many nests at once, recording what really goes on (nothing is going on most of the time at any given nest because brooding precludes mayhem, so an observer scans across the nests for signs of parents getting up). In both species, the key events of the siblicide process are predictable in time: the action will definitely take place within two or three days after hatching, and probably during daylight hours (because of nocturnal brooding). In a pelican colony, then, a biologist merely needs the patience to sit quietly for a few days, keeping an eye on several nests that hatch simultaneously. If there is much variation in the hatching dates, the observer may simply sit there for a few more days and get full second and third multi-nest samples during the same season. The numbers can accumulate rapidly.

Indeed, the main problem for the colony-based researcher is the inevitable disturbance associated with human traffic: entering and leaving the blind, marking nests and eggs, and so forth can upset parent birds and risks the possibility that they will desert their nests. And so biologists have devised various methods for minimizing disturbance, including the use of above-ground "tunnels" (black plastic draped over bent metal bars to make a human-scale molehill) and tricks for habituating adult birds to human presence. Hugh Drummond of the Universidad Nacional Autónoma de México has a great situation: the siblicidal boobies he studies breed on a remote island, Isla Isabela, surrounded by a clear tropical sea off Mexico's west coast. Because of their isolation, these birds exhibit very little fear of humans. With no need to conceal himself inside a blind, Hugh sits comfortably in a lawn chair beside whichever nests he wants to watch.

Scientists inevitably face serious methodological trade-offs when selecting a study system, though I continually hear colleagues at conferences announce that their particular research animal is "ideal" for exploring this or that. In fact every subject species has inherent disadvantages, but some systems offer better compromises than others. Accordingly, when students tell me they are keen to study tigers or whales or some other glamorous megavertebrate, I'm likely to launch into a short sermon on the pragmatics of a data-rich alternative. While the world certainly needs to know more about rare and exotic creatures, including tigers and whales, important general principles are more often learned from their commonplace and local counterparts. A fine example of such a research program is that of Nick Davies, who has spent many years studying the dunnock (a.k.a. hedge-sparrow), just outside his office on the Cambridge University campus.[12] This ultra-drab little brown songbird does many extraordinary things, the most famous being that the male stimulates his mating partner to expel her previous sexual partner's ejaculate by pecking her derrière just before mounting.

Getting back to the logic of egg insurance, we need a framework for understanding how natural selection deals with catastrophic risks to very young offspring. Let us imagine some probability p (a value between 0 and 1) that a given egg will fail to hatch. The probability that it will *not* fail but succeed in hatching is therefore $(1 - p)$. If parents lay only one egg, then, they face p chance of disappointment after going to the trouble of mating, laying, and incubating for many days. In high latitudes, this opportunity cost may easily be the whole breeding season, since each nesting cycle takes from weeks to months to complete.[13] The insurance coverage that parents stand to gain by producing a second egg can be calculated by looking at how their risk of total failure is affected. If we make two reasonable initial assumptions, (1) that a second egg is essentially a duplicate of the first, specifically that it has the same chance of hatching, and (2) that its presence does not lower prospects for the A-egg (that any possible effects of crowding or heat-sharing under the parent's breast are trivial), then adding a B-egg cuts the risk of total failure down to $p \times p$, the chance that *both* will die before hatching. Similarly, there are calculable probabilities that the A-chick will hatch and B will not, namely $(1 - p) \times p$, the same as B's chance of being the sole hatchling, $p \times (1 - p)$, and finally a chance that both will emerge safely from their shells $(1 - p)^2$.

To get a better feel for this argument, let us imagine that two out of every ten eggs ordinarily fail for various reasons (genetic disorders, pathogens, single-egg predation, accidental cracking, and so on), this being the ballpark value just determined for white pelicans and black eagles. (We have no figures to date for how many of the pelican and eagle A-sibs that perish never reach the point of hatching, but that is probably the typical pattern.) Substituting .2 for p in all the expressions above, we see that once every five breeding cycles parents opting to lay just a single egg would fail to get a hatchling, while those that made the additional investment in a second egg would push their risk of such catastrophe down to .04 (4 percent). In species like eagles and pelicans that nest just once per year, such insured parents will face that disappointment only once every quarter-century. Using these same values, we see that our two-egg ("insured") parents would end up hatching *only* their A-chick 16 percent of the time, *only* their B-chick 16 percent of the time, and *both* members of the two-chick brood 64 percent of the time. In effect, the investment associated with creating and incubating that second egg, the up-front premium on the insurance policy, cuts the parents' vulnerability to disaster to one-fifth of its one-egg risk.

But doing so also creates a new problem. In nearly two-thirds of the nests there will need to be some form of brood reduction later on, to cancel the insurance policy when coverage is no longer needed (to eliminate the "redundancy problem"). The costs associated with such culling may be considerable, for example if the victim can fight back effectively, but in obligate siblicide species they are kept low by parental favoritism. In particular, parents typically combine early incubation of the A-egg with a sizeable interval between it and the laying of B to create unusually long inter-hatch intervals. In fact, eagle species that practice obligate siblicide have longer intervals than other two-egg eagles, including species that practice siblicide only sometimes and frequently raise both chicks to independence.[14]

In principle, hatching failure rates could be so high as to favor *multi-*

ple backup eggs (the chance of complete failure being p^n for a clutch of n eggs), just as a football roster always carries third- and fourth-string quarterbacks as extra insurance against linebacker terrorism, but this seems not to be practiced by many (perhaps any) birds, plausibly because avian hatching success tends to be quite high. To finish off our hypothetical example with $p = .2$, one can see that a third egg would have much less impact on reducing calamity (to $p^3 = .008$) than the second egg, and a fourth egg would make a truly minuscule contribution (to $p^4 = .00016$). Besides, even just a third egg might complicate the redundancy problem in the hefty proportion of broods where all three succeed in hatching, $(1 - p)^3 = .512$, and double killing would be needed to trim brood size down to one. The brown pelican is the only bird I know of that has a three-egg clutch and may have an obligate type of siblicide,[15] but more data on the frequency of siblicide are needed for this species before we can be sure.

The way I have been building the case so far, readers can be forgiven for thinking that all birds lay at least two eggs, but that would be incorrect, and so it is worth considering the conditions under which laying of an insurance egg does *not* occur. It is always a bit dodgy to ask why something has *not* evolved, since one answer could be that the genetic basis for any given trait (in this case mutations impelling females to lay beyond the first egg) simply has not arisen by chance. If the genetic variation is not present, its effects cannot be adjudicated by the process of natural selection. So, in the following discussion there is the tacit assumption that such variation (alternative alleles promoting production of either one or two eggs) has been available in the history of the species being discussed.

Single-egg clutches are common in certain seabirds, eagles, pigeons, and even some songbirds. At 1,300 grams, Wahlberg's eagle is smallish for an eagle, but it offers the fascinating attribute of laying either one- or two-egg clutches, thereby inviting comparative study within the same population. That is, this bird may be thought of as sitting astride the life-history decision of which clutch size is better. Of course, there may be a subset of the population for which one egg is the optimum and another for which two has the bigger payoff, in which case we would want to know the correlates of each choice.[16] But too few pairs (2–3 percent in the most closely studied population) lay two eggs for much straight comparative work. If both eggs hatch, siblicide normally follows quickly. Acknowledging that the eventual survivor in two-egg nests is sometimes the B-chick, validating the potential for insurance, Rob Simmons focused on the question of why *all* pairs don't lay a backup egg.[17]

About one-eighth of all eggs in two sub-Saharan populations fail to hatch, so we can use $p = .125$ to generate our baseline estimate for the insurance potential. As explained earlier, a second egg might drop this failure rate down to $.125^2$ or .016. All else being equal, dropping the risk of wipeout from 12.5 percent to 1.6 percent looks like quite an attractive payoff.

An experimental approach was taken to see whether parents could, in fact, afford to raise two chicks. In some obligate siblicidal species, such as the black eagle, even massive efforts by humans to keep the chicks from killing one another (hand-raising each one in captivity for several days, then swapping them), interspersed with periodic tests (placing the chicks together in the nest to see if their aggressiveness has subsided) have thus far proven futile.[18] No quarter is given, ever. But in Wahlberg's eagle such a truce has been determined to commence about thirty-six days after hatching. Thus, eight pairs of Wahlberg's eagle parents with single seven-week-old chicks received an extra mouth to feed, a similar-aged second nestling. It was not attacked by the resident chick, which was good for the study's aims, but two chicks turned out to be more than the adults could (or would) support. Limited observational records from two of these nests revealed apparent harmony between parents and the new chick, but very little food delivered. The slightly smaller chick in each nest, which was sometimes the original resident and sometimes the introduced one, lost condition and died in all but one case, and even that lone exception was 20 percent underweight at fledging. Furthermore, the parents that had maintained two-chick broods for at least ten days turned out to be less likely to breed the following year, so some wear-and-tear costs, presumably reducing parental fitness, are indicated as a disincentive for trying to rear two chicks.

Before I discuss Rob Simmons's interesting conclusions from this project, I wish to note that these data are nicely consistent with the notion of "pending competition" that some raptor biologists have resisted. At the very early stages, when siblicide normally occurs in this species, parents presumably are able to provide more than enough food for both tiny chicks, but we see from these data that such may not be the case later on. Also, the fact that fooling parents into bearing the double load for just ten days impairs their future reproductive performance strongly implies that parental interests may be served by early sibling aggression and fatality in two-chick broods: such parents may be aiming for one chick all along. But this logic is slightly forced since these particular parents all had produced just one egg in the first place: it is perhaps unsurprising that their game is spoiled by an imposed experimental burden. It would be nice to know, in fact, if these parents actually worked any harder *than usual.* They certainly did not bring home much food (one small reptile weighing 50–100 grams every 6.7 hours of observation).

Simmons took a hard look at egg size in single-egg clutches and found a strong relationship between an egg's volume and its probability of hatching, with all big eggs (> $76cm^3$) hatching but smaller eggs (< $70cm^3$) having only a fifty-fifty chance. Under the reasonable assumption that a finite supply of egg material can be packed into either one big egg or two somewhat smaller ones, even correcting for the possibility that the mother would have a few extra days to metabolize food under the two-egg option, a minimum size for hatchability could tip the balance in favor of building just one. Furthermore, across the eagles that practice obligate siblicide, B-eggs tend to be about 12 percent smaller, which may reflect a compromise skew, wherein the more valuable A-egg is given first class treatment while the B-egg is assigned a seat far back in coach. In other species, it has been proposed that the insurance egg ought to be as cheap as possible without losing its viability.[19] So egg quality may need to be factored into calculations of insurance value and, in the case of Wahlberg's eagle, this consideration seems to favor building a single, large egg. Interestingly, in the few nests where mothers laid two, A-eggs were huge and the B-eggs were also large, but variable. I find Simmons's interpretation that insurance is not the primary selective pressure behind producing a second egg in this population to be a bit off the mark, but it seems that an alternative way of insuring hatchability (by increasing egg volume) may have usurped the potential role that a second egg might serve for most pairs. For the few parents that can afford both a large A-egg and a backup B-egg, and especially for a more equatorial population that is reported to have a higher incidence (10 percent) of two-egg clutches, we must still wonder.

Some parallels exist in another raptor, the Hawaiian hawk. These mothers also lay very large eggs (roughly 10 percent of the mother's body mass), but only one per nesting cycle. From the published egg-mortality and chick-mortality data,[20] it appears that having insurance offspring would be valuable because there is only a fifty-fifty chance of a freshly laid egg surviving both its lengthy (38 day) incubation and even longer (59–63 day) nestling periods. So enormous egg size is unlikely to be the whole story. Instead, the lack of a backup egg may be related more to opportunities for pairs to start over again quickly if their sole offspring dies.[21] That is, the option of re-laying may be much more feasible for this sedentary, tropical raptor (which can breed year-round) than for the migratory, and hence seasonal, Wahlberg's eagle. But even that eagle is able to replace eggs quite rapidly: if the single egg is stolen early in the incubation period, a replacement can be produced within just 9–11 days, an option that may further relax the incentive for insuring with a second egg.

Another reason for birds to pass up the option of creating an insurance egg concerns the early costs of buying and retaining the policy throughout the period of risk, a point I finessed earlier by noting that a black eagle egg is only 2–2½ percent of its mother's body mass and implying that such a small value is inconsequential. In fact, until very recently nearly everyone contemplating such questions was under the impression that production of one more egg was likely to be nearly trivial. But some eggs are much larger on a relative scale (some seabirds lay eggs that are 25 percent of maternal mass),[22] and new evidence is emerging that building and incubating an extra egg can involve serious costs, even for birds with "ordinary"-sized eggs. Clearly we shall need to know more about these debits for many species before we can draw firm conclusions.

The newer findings, coming out of Patricia Monaghan's group at the University of Glasgow, are based on simple experimental manipulations of parental effort. From one sample of nests in a colony of lesser black-backed gulls, the A-egg was removed the day it was laid, as if stolen by a fox. This act forced the mother to lay a total of four eggs in order to reach her rigidly set goal of three (such numerically goal-oriented birds are said to be "determinate" layers). Then the parental performance was evaluated, along with physical impact on the mothers, relative to a similar sample of control families whose mothers were not pressed into producing an extra egg. In both treatments the parents incubated the same number of eggs (three) and faced otherwise similar conditions. But in laying that fourth egg, experimental mothers had to mobilize materials from their large pectoral muscles, the primary muscles of flight. Fewer of their chicks survived the nestling period, and the ones that did manage to fledge weighed less, a common correlate of lower post-fledging survival in gulls and many other birds.[23]

To see if a bit of extra incubation also imposes measurable stress on parents, the biologists experimented with common terns, another graceful coastal bird that nests in dense colonies. These terns may lay one, two, or three eggs, so the researchers focused on twenty-one pairs that had stopped with two, transporting a third egg to these clutches just two days after laying had ended (as soon as it was clear which pairs were not going to lay a third egg of their own). The only differences between these experimental pairs and a parallel set of twenty controls (which laid two eggs and received no extra one) were those associated with incubation itself, because the introduced third eggs were returned to their home nests just prior to hatching. Here, again, the chicks of taxed parents fared less well than control chicks, with an interesting twist that we shall see more of in a later chapter. The slight (one-day) hatching delay experienced by B-chicks put them at such a competitive disadvantage that poor food deliveries by experimental parents led mainly to deprivation for B-chicks, causing them to grow more slowly and fledge at lighter weights than control B-chicks. By contrast, experimental A-chicks matched their control counterparts well.[24]

The implication of these studies for insurance eggs concerns affordability. Building or caring for one extra egg may be a good bit more expensive, at least for some birds, than we first realized. If this turns out to be generally true, it may help explain why more species have not evolved the practice of creating a backup offspring.

6

Killing Me Softly

When any relationship is characterized by difference, particularly a disparity in power, there remains a tendency to model it on the parent-child-relationship. Even protectiveness and benevolence toward the poor, toward minorities, and especially toward women have involved equating them with children.

—Mary Catherine Bateson, 1989

Lethal aggression against siblings certainly seems an expensive way of reducing competition for limited local resources. There is the physical exertion to consider (remember the 1,569 eaglet pecks?) and also the potential for retaliation and injury to the attacker. Though the dangers from effective counter-punching are probably not much for eagle and pelican A-chicks, who possess huge size advantages conferred upon them by hatching much earlier than the victim sibs, their zeal for getting the job done as quickly as possible may be advantageous in part because the task of execution is likely to become more difficult with the passage of time. Delay probably also adds risk of a *coup d'état* by allowing a would-be usurper to persist, especially if the incumbent ever becomes temporarily weakened (and exposure to food-borne pathogens may well impose an ironic penalty for doing most of the eating).

In many systems, sibling rivalry takes various nonaggressive forms that have the same fatal effect of permanently removing a within-nursery competitor. On the campus of the University of Bangalore in India is a tree, the Indian rosewood, that goes by the Latin name of *Dalbergia sissoo*. The sex life of *Dalbergia* begins and ends when pollen reaches the flowers of the mother (the tree or "maternal sporophyte"). Pollen grains are tiny but rugged travel cases that contain two sperm cells of identical genetic composition (having split mitotically from a single sperm), so they have no intra-pollen conflicts of interest. Upon reaching a female flower, one sperm grows a long tube into the ovule while the other divides once more before the (new pair of) twins travel down the tube to the ovum and fertilize. You may have noticed here that two male gametes are involved but just one female gamete: it is actually a bit more complex than that, but we can defer that business till later. The *Dalbergia* nursery is a long, oblong pod, rather like those in which snow peas reside. The original flower, out at the pod's distal tip, falls off once fertilization has been completed. The linear arrangement of the four to six sibling seeds is what interests us, because the distal ovum gets fertilized first, even as rival pollen teams are growing their necessarily longer tubes to reach the more distant ova. Like an eagle A-chick, this distal seed thus obtains a temporal advantage over its not-yet-fertilized pod-mates.

The just-activated distal seed's physiology kicks in immediately with a rapid growth program that generates some water-soluble chemicals that pass inward toward the stem (that is, toward the tree's trunk), passing the sibling seeds en route. This dissolved chemical bath kills virtually all of them over the next few weeks. Thus the number of seeds per pod can be as high as six just after fertilization (the mean is 4.25), but usually drops to just one (mean 1.2) by week six.[1] Simple but conclusive laboratory experiments have demonstrated the active role played by the "dominant" distal seed. Living seeds placed on standard agar blocks were treated with either plain water, extract from aborted (dead) seeds dissolved in water, dissolved extract from healthy seeds, or dissolved extract from pod tissue (that is, essence of mother, not of sibling). The extract from healthy seeds caused double the mortality of the other three treatments, which were all similar to one another. This shows that the killing is due to the dominant sib's physiological activities and not to those of the mother. It also demonstrates that the victorious seed has to be metabolically viable.

Occasionally, this chemical sabotage of siblings does not work properly and up to three seeds may survive in a single pod. When this occurs, the distal seed is smaller than it otherwise would have been. As well, the whole pod is then heavier than if the culling process had functioned normally. As in many plants, offspring success requires a dispersal step, wherein the pod detaches from its branch and is borne by the wind to a new home on the soil. Heavier pods have a higher wing loading, which presumably affects how far the ripe seed disperses from the maternal seed shadow. A longer flight may improve its chance of landing on a suitably uncrowded growth site, farther both from its well-established mother and from where her other offspring will land. Finally, if the distal seed in a pod dies for some reason (including being snipped out with a pair of scissors by a curious biologist), the next-most-distal seed takes the dominant role: on average, one-third of all pods end up carrying an initially nondistal seed. The insurance value of making extra seeds seems apparent.

We turn next to a mammalian parallel involving obligate mortality without fighting, namely the pronghorn, a North American animal known colloquially as the pronghorn antelope. The recently mated female may have a half-dozen or so tiny "blastocysts" (very early sibling embryos that have taken on a globular structure with walls just one cell thick) drifting inside her like water balloons in a pond. As development continues, these grow into stretched and skinny tubes that can reach twelve centimeters in length. The first wave of offspring losses occurs when some of these become fatally entangled with one another. Only a few survive to the implantation stage, when each must attach itself to a permanent uterine site where the placental connection will develop.

The pronghorn uterus is "bicornuate," meaning that it has two arms or "horns," and within each horn there are two implantation sites, one nearer the birth canal and another slightly deeper back. In the early stages of pregnancy, the blastocysts float freely back and forth between the horns for a few weeks, with losses occurring anywhere in that space. Eventually, though, about three thread-stage embryos find the attachment sites and hunker down, with two in one horn and the third by itself. Soon their placentas develop and the more efficient transfer of nutrients from the mother's metabolism to the embryos allows development to accelerate. The rapidly growing embryos are oriented with their front ends facing the birth canal. Their fronts remain rounded, but their tails elongate and become pointy. Furthermore, the two implantation sites within each uterine horn are not of equal quality, with the one in front (closer to the birth canal) having a richer capillary bed. This confers a growth advantage on the front offspring, eventually bringing about a meeting of its pointy tail with the face of its trailing sibling. The fatal skewering of the victim that follows is due simply to continued growth of the tail (colorfully nicknamed the "necrotic tip" by its discoverer, Bart O'Gara) through the full length of the sibling embryo's body.[2]

Final litter size for pronghorn neonates is thus two. It seems plausible that the second implantation site may provide one of these survivors in the event that the sibling up front fails to develop properly, and so may be yet another case of marginal offspring as insurance, but no data exist on this key point.

In this age of modern medicine, we hear increasingly of women being artificially stimulated with hormones to release multiple ova and finding themselves pregnant with litters of nonidentical ("fraternal twin") embryos. Presumably, this has been going on in pronghorns for millions of years, and it turns out that polyovulation is well developed in four unrelated mammalian groups, the other three being the elephant shrews of Africa and Madagascar, a peculiar South American rodent called the plains viscacha, and certain bats. The record for initial clutch size seems to belong to the plains viscacha, an enormous (some males exceeding twenty-four pounds), colony-dwelling Argentine chinchilla, which has been found to have 845 ova in a single female.[3] All of these very different species produce tiny litters (just one offspring per uterine horn) of large and precocial young. In addition, all feature a maternal lifestyle of high mobility. Elmer Birney and Donna Baird suggested that the mothers' need for agility in escaping predators may have trimmed optimal litter size to two and that the initial overproduction above that number is likely to have evolved for insurance benefits.[4] Why it skyrocketed into the hundreds in the viscacha remains a bit of a riddle (the others are considerably more modest), unless the ova are so cheap as to provide little in the way of a constraint.

Now let us return to birds whose nestlings are relentlessly destructive (of which we met several types in Chapter 5) for two examples of obligate brood reduction *sans* fighting. Both involve penguins, and in both the parents are playing some unusual behavioral and life-history games with their small two-chick broods. Dee Boersma, a conservation biologist at the University of Washington, has spent much of the past twenty years at Punta Tomba, a beautiful and isolated bulge on the southeastern coast of Argentina, where she has kept track of thousands of individual Magellanic penguins, a population in sharp decline. When one is concerned with dwindling numbers of a species, the automatic loss of half of each season's offspring soon after hatching calls attention to itself. It seems that Magellanics lay two eggs but raise only one chick, plausibly because one per year is all the parents can afford, with the second playing the now-familiar insurance role. What is interesting in this system is the direct parental role in skewing food allocation toward the A-chick. In nests where both eggs hatch, the parents simply offer no sustenance to the B-chick, turning again and again toward the A-chick when regurgitating food.[5] As a result, the B-chick quickly starves to death. So we see that the early parental act of incubation during laying, which produces hatching asynchrony, is followed with overt behavioral favoritism toward the offspring that already benefited from the parentally bequeathed head start. This is what I meant when I proposed parental complicity.

The crested penguins throw another wrinkle into the story (probably several wrinkles). These five or six sub-Antarctic species all sport rakish yellow plumes above their eyes and exhibit the most lopsided division of egg materials known among birds, with the smaller egg having only 15–42 percent the volume of its larger nestmate.[6] Stranger still, this time it is the first-laid egg that gets short shrift. This size asymmetry might all wash out if parents began incubation right away and gave the small-egg embryo a compensating temporal advantage, as it takes another four to five days to construct the big egg, but they do nothing of the sort. Instead, incubation begins *after* the second egg is laid, at which time *that* egg is tucked into the safer, rear position beneath the linear brood patch of whichever parent is incubating.[7] The chick from the big egg hatches either on the same day as its smaller sib or even a day ahead.[8]

Thus equipped with nutritional and sometimes with temporal advantages, the chick from the second-laid egg generally proves to be the sole survivor. And instead of using aggression to intimidate or getting conspicuously preferential parental treatment, the victor simply outreaches the victim. It is so much heavier that it can rump its way into position and so much taller that the parent's descending bill reaches it first. By

Crested penguins lay two eggs of different sizes. The second-laid egg is both larger and usually the first to hatch. If both chicks hatch, the smaller one is not physically abused, but it usually dies early from being unable to compete with its stronger sibling for food. (Photo: Timothy C. Lamey.)

the time the chicks move into large groups known as crèches, when they are approximately a quarter of the way through the parental investment period, nearly all broods have dropped from two chicks to one. The crèching youngsters huddle together, largely for safety in numbers, separating from the group only when visited by food-bearing parents (who deliver food only to their own chick). About six-sevenths of these survivors came from the larger egg in their home nest.[9]

I have had the pleasure of serving as advisor for two Ph.D. students, Tim Lamey and Colleen St. Clair, who sallied forth to study crested penguins under the chilly and extraordinarily windy field conditions of sub-Antarctic islands. They learned many fascinating things about how the drama plays out, who survives, and so on, but the central riddle lingers as to why it is the second egg that gets the sweet deal, when all other birds (including other penguins) that have asynchronous hatching and brood reduction favor their first egg. Some force(s) in the distant past apparently tipped the balance, quite possibly only once, setting the historical inertia toward this odd pattern. This presumably involved a chance genetic novelty occurring when some ecological conditions made such an arrangement advantageous, but we can only speculate about what those conditions might have been. For example, first eggs may have been unusually vulnerable to predation for some reason.

Arguably the most bizarre of the crested penguins is the royal penguin. Here the A-egg suffers all the usual indignities for this genus (small size and lack of early incubation), but then something amazing happens: just before the mother lays her big second egg, she kicks the first one completely out of the nest, such that it is never incubated and its resident embryo dies.[10] To document this extraordinary behavior, Colleen and her husband Bob had to sit quietly for days, facing into ferocious and frigid winds—because the penguins, whose bellies they needed to be watching, choose to stand over the nest scrapes with their backs, quite sensibly, to the wind.

Trying to figure out the biological significance of parents going to the trouble and expense of creating an egg—even a small egg (and the discarded egg of royal penguins is among the smallest of all crested penguin eggs)—only to boot it out without giving it a real shot at survival is enough to make one want to join the "inexplicable biological waste" crowd. Some have proposed that the crested penguins in general, and royal penguins in particular, represent an evolutionary transition, such that the smaller egg is simply in the process of being phased out (like middle-management jobs in a downsizing corporation). I suppose this could be so, but I am uncomfortable with it for two reasons. First, shrinking one of the eggs seems like a strange way to make a clutch-size adjustment. There might be some point at which further size reductions would utterly compromise viability, at which point its production would become maladaptive. But if the small egg currently has *no* reproductive value, apparently not ever, why not divert all egg materials to the winner, as most Wahlberg's eagle pairs seem to do? Second, the evolutionary transition argument strikes me as a logical dead end, a facile and unproductive dismissal of some problem we simply do not understand yet. A few decades back one could have found behaviorists who would have explained the maternal ejection by announcing simply that "it must be an instinct," as if pigeon-holing it into a named category was equivalent to understanding it.

My preference would be to try coming at the riddle from some new directions and keep testing possibilities. I freely acknowledge that evolutionary transitions must exist, but it strikes me as an unlikely coincidence that Charles Darwin should have happened to come along and get us thinking about *why* questions at just the moment (a century or two being just an eye-blink on the evolutionary time scale) that all half-dozen of the crested penguins simultaneously chanced to be in the process of putting their ancestral two-egg clutches to rest. Thus, my intuition leads me to regard the currently observed patterns as more likely to be evolutionarily stable equilibria. If I could get a grant to visit the Pleistocene, I would expect to see crested penguin ancestors laying one small egg and then one big egg, much as they do today. If so, then this pattern may simply represent a mixed balance of costs and benefits that we should be able to identify and measure.

Therefore, while acknowledging that the null hypothesis could be correct (that these reproductive features of crested penguins could have no significant impact on parental fitness), in which case no further explanation of current function would be needed or possible, the very oddness of the traits suggests otherwise to me. Egg size is well known to affect offspring fitness in birds with milder variation, where it has been studied more closely over many seasons: in many cases, chicks from bigger eggs are more likely to survive and become successful breeders themselves. So why should egg size be neutral in the one genus whose intra-clutch variation is an order of magnitude greater than all the rest? And, of course, why should the *first* egg be the marginal player in this group, when other brood-reducing species almost universally assign that role to the last-laid egg?

The practice of asking "why" questions about interesting phenotypic traits in terms of their current costs and benefits is the basic approach of behavioral ecologists, but it is not without its detractors. One common complaint is based on the argument that natural selection is not directly responsible for all features of every organism's body, that there are several strong forces (genetic mutation, gene flow between populations, each trait's historical past and how that phylogeny limits what can evolve next, and so on). This position was popularized by Stephen Jay Gould, who attempted to paint "adaptationists" into an extreme corner, likening them to orthodox converts to Rudyard Kipling's famous *Just So Stories.* According to Gould's parody of the approach, adaptationists make up a single, vaguely plausible story that might account for something (such as How the Leopard Got Its Spots) and then they slap themselves on the back in congratulations and all go home. Gould and his colleague Richard Lewontin also opined that adaptationists believe that the process of natural selection, working alone, routinely polishes every feature of each species's life into a gem of perfection. This caricature was packaged with a clever nod to Voltaire's Dr. Pangloss, who viewed the world as the best that could possibly exist. An orientation toward asking functional questions in biology, thus, was labeled the "Panglossian paradigm,"[11] with the explicit accusation that scientists who turn their attention first to the possible role of selection believe that every detail of every trait is, in essence, perfect.

This is sheer nonsense. Virtually all scientific explanations are tentative, based on incomplete information and therefore subject to later revision and/or rejection. Such conjecture is not unique to behavioral ecology, nor to biologists interested in the actions of natural selection. When we begin to understand some portions of a biological phenomenon, a reasonable next step is to propose some preliminary explanations for why it may have evolved as it has. If those hypotheses lead to testable predictions, they are useful: if they are either untestable or foolishly enthroned as the final word, they are likely to be ignored. My point is that a just-so story accommodating all the currently available information on its subject is a necessary part of the process. Most important, it is a transient step. For example, Collingwood Ingram's nice story about the short-eared owl larders (see Chapter 5) may have been set out more strongly than the evidence justified, but the story did no lasting harm by itself. And it may yet prove close to the truth. I noted its thin empirical support to show how general impressions (in that case about sibling cannibalism) sometimes tend to build up. But his hypothesis is still an interesting one and potentially quite testable.

Anyway, if I am to be viewed by some as a Panglossian adaptationist, I may as well finish this chapter by going back to those drop-kicking royal penguin mothers and offering an insurance-based argument, in the hope that someone willing to go off to the sub-Antarctic may find my musings worth testing. If I am dead wrong, so be it; that comes with the job. Here's my hunch: perhaps the only significant risk faced by the mother in this particular breeding mode lies in an occasional inability to mobilize resources for her one big egg. If so, then laying a tiny first egg may eliminate most of the chance of *complete* failure for the cycle, by providing a cheap backup. However, once the large egg has been built and is safely in her oviduct receiving layer upon layer of nutritive materials (the albumen), the insurance policy may no longer be needed (and, as we have seen, incubating is probably not cost-free). The evictions that have been observed were all performed by the mother herself, none by the father, and most occurred during the twenty-four-hour period just before the laying of the big egg.[12] By then, the mother's physiology may have solid information that the favored chick is going to make it. Well, that's my thought on this problem. I look forward to reading what the next batch of intrepid penguin biologists can learn.

7

Parenting in an Uncertain World

A little more than kin, and less than kind.

—William Shakespeare

The last two chapters were all about family systems in which disposing of nursery-mates—either a single sibling (the bird and pronghorn examples) or many siblings (*Dalbergia* and the plains viscacha)—is the rule, rather than the exception. In such families, parental overproduction is routine (especially to gain marginal offspring as backups) and secondary culling is essential. When the risk to the survival of core members is either negligible (for example, if producing larger eggs enhances hatching success) or easily repaired (by starting over again immediately), the insurance value of overproduction ebbs. Once you get used to it, the idea of parents making surplus offspring, even in species where parents must "know" in some sense that not everybody is likely to make it, makes intuitive sense. The problem is that parents cannot always identify *which* kids they will lose.

The next step in exploring the biology of family dynamics is to relax the assumption that brood reduction is inevitable. How does the game change when resources frequently do prove to be adequate? We turn to the broad subject of incomplete information. In most plants and animals, parents probably cannot foresee how many offspring they will be able to afford for their next reproductive cycle. This is because the family income is impossible to predict over the medium and long term (and sometimes even over the very short term). For example, as I write this in late January, I know that today's forecast here in Oklahoma is for sunny skies and a high of about 40 degrees Fahrenheit. I recall that yesterday's forecast (similar, but slightly warmer) was accurate, so I shall dress for today's predicted conditions. With weather satellites hovering high over the planet these days, those of us living in the Great Plains can often be given a pretty good idea of what tomorrow's weather will be, too, because most storm fronts are visible to modern technology while still a couple of states away. But the stakes for making accurate predictions (or, lacking that, for taking conservative precautions) were higher for our ancestors, just as they still are for animals other than humans.

If you come back in the next life as a cattle egret, your first few weeks will be spent in a modest stick nest, roughly the diameter of a basketball player's handspan, with a couple siblings and an unknowable family income that may or may not require someone to die. Your parents will make their living by catching grasshoppers between two forceps-like mandibles. It's not an easy job. For one thing, natural selection has tuned the DNA in grasshopper cells to build bodies good at avoiding such a fate. Grasshopper DNA produces grass-colored cuticles and the insects' talent for freezing on a stem and pretending they don't exist whenever a large being approaches. A cattle egret hunting alone probably walks right past most of the grasshoppers it encounters. But cattle egrets long ago made the important discovery that grasshoppers cannot afford to bluff a rhino. With a behemoth bearing down on its hiding place, not caring in the slightest whether it looks like vegetation, the grasshopper must adopt Plan B, leaping from the path of destruction before freezing and pretending not to exist once again. A cattle egret walking at the head of a grazing rhino, zebra, or cow is nothing if not alert to the flicker of such motion. It fixates on the spot where the insect lands. And, of course, it is a lot better at grabbing things with forceps than you and I will ever be, at least in *this* life.

An obvious problem arises for cattle egrets whenever weather conditions change so as to depress grasshopper (and ungulate, for that matter) activity. If it is too hot, too cold, or too rainy, everybody gets sluggish. Inactivity by the other players cuts the egret family income sharply, and if the downturn lasts more than a couple of days, the resource shortage may create a fatal squeeze on family size. To my knowledge, nobody has checked whether egret foraging effort increases as a storm approaches; some birds are sensitive to barometric pressure shifts (important when a migration path crosses a wide body of water like the

An adult cattle egret perches on a cow's back during a break from its foraging. Cattle egrets hunt just in front of large grazing animals, capturing grasshoppers that jump to avoid the ungulate's hooves. (Photo: Douglas Mock.)

Gulf of Mexico), so a bit of parental hustling before a storm does not seem too far-fetched.

Parallel problems affect great egrets, a taller cousin that practices the more typical egret habit of wading in shallow water in search of small fish and frogs. Just as cattle egrets are adept at grasshopper hunting, great egrets are terrific fish-grabbers. The two species share some physical specializations for spear-hunting, including, for example, densely packed receptors positioned in just the right spots on their retinas to deliver exceptional ability to focus on small objects straight in front of their heads. Another anatomical specialization is an elongated sixth cervical vertebra rigged with tendons and muscles that allow the bill to be slung forward faster than prey can zigzag. Skilled predators though these egrets may be, they live near the margins of their budgets. So, when rainy weather dimples the pond surface or keeps grasshoppers under cover, the family income can take a hit. And rain, of course, is not the only problem affecting this budget. All kinds of biological headaches can either reduce prey populations (viruses, bacteria, predators) or make them less available (as when other predators make them wary).

If and when food supplies begin to fall, the continued existence of all the nestlings may become a luxury the family can no longer afford. Under other conditions, when the food base remains comfortably above any such dangers, the chicks are more valuable alive than dead, even to their nestmates, since a live sibling may go on to produce nieces and nephews. In between these extremes is an iffy world of gray tones, where food may or may not suffice to meet the needs of all family members and some hard choices must be made, often well in advance of the actual food shortage. Creating contingency options may be the best one can do.

Many parent birds establish such options by bequeathing what seem to us as unfair advantages to some offspring and/or disadvantages to others. The resulting asymmetries impose a social hierarchy on the nestmates, which probably helps hedge parental bets against environmental vicissitudes. The first of these to be considered by ornithologists was asynchronous hatching of the clutch.[1] Parents have simple ways of engineering broods that range from being physically matched for size to highly disparate (just how they pull this off will be addressed in Chapter 9). The trick lies in parents controlling when each embryo starts development and, as we have seen, siblicidal eagles, pelicans, and other species can give head starts of two days or more to A-chicks. On a more moderate scale are most songbirds (the vast order Passeriformes, to which roughly half of all birds belong), where the typical pattern has all but one chick hatching within a few hours and the laggard emerging from its shell on the following day. Unlike the pugnacious birds discussed so far, songbirds are essentially pacifists, using little or no overt aggression in settling their disputes over food.[2] One could fret that perhaps it simply has not been observed yet, but songbird nestlings have been subjects in hundreds of field studies in which the young are visited and handled on a regular basis, so it seems unlikely that sib-inflicted wounds would have escaped attention. Moreover, one can imagine good reasons why many small birds are not physically abusive to one another. For one thing, most lack the kind of bill anatomy useful for meting out injury and pain. The one passerine known to be siblicidal, the black-billed magpie, possesses a tiny hook at the beak's tip, which normally helps in the shredding of small vertebrate prey into bite-sized chunks. It seems that when times are lean (cold, rainy), some senior sibs in a Montana population turn on their nestmates and use the weapon on them.[3]

Another group of hot candidates for passerine pugnacity would have to be the shrikes. These predatory songbirds have hooks with a sadistic little falcon-like notch near the bill's tip (a feature that adults apparently use for snapping the spines of snakes, lizards, and other birds, especially sparrows, on which they prey). Adult shrikes also kill prey in advance, hanging corpses around the territory on any suitable spike they can find, be that a thorn, cactus spine, or, in recent decades, a prong of barbed wire. I understand from a shrike maven that their nestlings show no signs of siblicide, but I remain open on this point. After all, the whole topic of siblicide was ignored for 2,300 years after Aristotle mentioned it, so perhaps a bit more scrutiny will change the picture (and there are many species of shrikes around the world that have yet to be studied). Then again, the adults' habit of stockpiling prey (shrikes are called "butcher-birds" in many places) may reduce the uncertainty of food deliveries to the young, allowing a cleaner match between family size and the food base. We need to know more about the significance of shrike (and short-eared owl) bank accounts.

Songbirds, though not generally siblicidal, do exhibit considerable levels of fatal sibling rivalry of a more genteel sort. Indeed the general term for mortality of certain but not all nestmates, "brood reduction," was coined for one such bird, the curved-bill thrasher. Back in the summer of 1964, Bob Ricklefs did a small undergraduate research project with thrashers in Arizona to demonstrate the phenomena underlying what I have been calling the resource-tracking strategy. Bob found that the food requirements of a one-chick brood did not overtax its parents' capacity to deliver, but broods of two or more did so.[4] In one three-chick brood, the early starvation of one nestmate relieved much of the family's supply-demand problem and resulted in only slightly retarded growth for the two survivors. In another brood, where four chicks hatched, the two youngest starved within a few days. For a while, the smaller of the two survivors continued to show retarded growth. Then an accidental fall eliminated its remaining nestmate, bringing the undersized nestling into a brief period of affluence, during which it enjoyed a growth spurt. But alas the whole nest tipped over a day later, spelling the last chick's doom. These thrasher anecdotes showed the kind of relationship that might exist between falling family size and growth patterns, which have been documented on a grander scale with various species since.

More typically, prolonged food shortages produce the death of just one offspring in a songbird nest. The loss of a single hungry mouth is often sufficient to ease the competitive pressure, thus sparing the rest of the brood. Early death comes most often to the last-hatched chick, the smallest and least competitive inhabitant of the nursery, which dies of starvation without having been physically harmed by its nestmates. So even in these pacifist species survival is problematic for the one marginal chick, the laggard that hatched a day after its nestmates.

We must consider why the degree of hatching asynchrony varies across avian families. What accounts for the differences between the highly exaggerated sibling mismatches found in the most inflexible brood-reducers (eagles, pelicans, crested penguins) and the very mild asymmetries of songbirds? But first we should take a moment to appreciate numerous bird lineages (such as ducks, turkeys, quail) where parents create virtually simultaneous hatching of even very large broods (often ten or so chicks). This group's offspring are over on the precocial end of the developmental spectrum, meaning that they are more like brush-turkeys than eagles. In particular, the chicks feed themselves. With the whole family's food not having to pass through the bottleneck of parental delivery, sibling competition is automatically relaxed. Selfishness, thus, is less important and so a higher degree of cooperation can be expected. In a few such species, laboratory studies have shown that the sibs communicate actively with one another while still inside their eggs.[5] It has been shown that these signals may help coordinate an even finer degree of brood hatching synchrony, presumably because that bit of timing allows the whole, ambulatory brood to vacate the vulnerable nest site en masse.

Getting back to the asynchronous hatching groups, there is as yet no sweeping explanation for the diversity observed, but I suspect that two life-history features account for much of it. Relative to eagles and their ilk, songbirds have very short life spans and nesting cycles. In a shorter life, each round of breeding constitutes a larger proportion of lifetime reproduction and is therefore worthy of something closer to all-out parental effort. Besides, those pulses of peak effort are short sprints, each round of breeding taking scarcely a month in many species. For example, the house sparrows that Trish Schwagmeyer and I currently study in central Oklahoma may raise three, and even four, broods in the March-to-August nesting period. Northern cardinals in this same area have been reported to manage five.[6] But adult sparrows and cardinals usually only live three or four years at the most. By contrast, eagle and egret nesting cycles last three to four months, occurring no more than once a year (perhaps less often), and adults can live for two decades or more. If an egret breeding effort fails, there may be too little time to start over, at least in temperate latitudes, forcing postponement until the next summer. By contrast, a backyard songbird whose nest is threatened by the local cat may spend an hour or two trying deterrence through dive-bombing, but it has a feasible backup strategy of quick and rather cheap re-nesting. If a cold snap cuts the flow of food, it can use the same backup. But egrets and eagles cannot play that speed game, so their better bet may involve incremental trimming of current family size when conditions are bad. And stepwise brood reduction is typically facilitated by asynchronous hatching.

We turn thus to species in which lesser siblings may have to die as conditions worsen, but do not have to die if conditions are good. In particular, we shall renew our acquaintance with the two egrets we have already met, both of which are said to practice "facultative siblicide," which means that the fatal assaults that seemed so inflexible in the black eagles and white pelicans (obligate siblicide species) are now subject to second thoughts and the frequent option of granting clemency. These egrets are still rough customers, with bullies that use aggression to establish and maintain domination over the designated victim. But the combat commonly stops at sublethal levels. Then we'll meet two other birds that forgo the violence but not the fatal consequences. Of these, one is a pacifist cousin of the egrets, the great blue heron, and the other is a ubiquitous songbird, the red-winged blackbird.

8

The Ultimate Food Fight

For every complex problem there is a simple answer.
And it is wrong.

—H. L. Mencken

Egrets (and most other forms of life on Earth) live in a world where food may or may not be adequate. The smallish cattle egret has a nesting cycle of nine to ten weeks ($3\frac{1}{2}$ incubating, the rest feeding ravenous nestlings), while great egrets may take another month or so. Given that it is not possible to know the food supply a month or two in advance, parents gamble by starting with more offspring than they may be able to support. Most successful pairs raise just two chicks to fledging, but they begin with three or four eggs and hatch them a day or two apart, thus over a span of three to five days for the whole set.

Because the three-chick brood is most typical in the populations I've studied, my research has focused almost exclusively on trios. Jennifer Gieg, a graduate student working with me, has shown that the asynchronous hatching is the result of the parents' commencing incubation between the laying of the A- and B-eggs. Jennifer placed a temperature-recording device inside a cattle egret egg that had been blown empty and refilled with silicone, then swapped that egg for the true A-egg, and recorded the temperatures it experienced during the nesting cycle's first few days. Comparing this information with that produced by a similar sensor-bearing egg in an unoccupied nest (no parents to provide extrinsic warmth) revealed what had been long suspected, that it is parental actions that raise egg temperatures high enough for the embryos to develop. The data show this boost to be inconstant at first (when only the A-egg is in the nest), but to become nearly continuous once the B-egg has joined the party. Thus A and B are given substantial head starts over the not-yet-laid C-egg. It has been argued that parents may have little choice about this early application of heat (that the eggs must be covered simply for protection),[1] but the parents' actions clearly affect the competitive composition of their brood. As we shall see later, there is good reason to believe that this parental manipulation of egret chick competitive asymmetries works in the parents' best interests.

The social struggles following hatching of the brood are infinitely variable in their details, though the broad picture is fairly simple. The A-chick usually picks a few fights in the early days, wins nearly all of those to reach a quick understanding with the other two sibs, then retires from the ring with nary a bruise or scar. It lives a relatively cosseted life in the weeks that follow, awakening when a parent returns from hunting with a meal. At such moments, it stirs and helps itself to more than its share (generally eating the first few boluses), without many obvious causes for worry. At the other end of the sibship, the C-chick's life is far more complicated. It gets only half as much food as its elders do, and it is an unwilling participant in (and loser of) the great majority of all sibling fights. In the middle position, the B-chick's role is pivotal. Strategically, it usually cannot challenge A for the free pass through nestling life, nor can it allow C to get strong and pose a potential challenge. If there might be a food shortage during the next month or two, it makes sure that C is the one to take the fall. In a nutshell, the real contest is for second place.

Not surprisingly, then, most fights in egret broods are between B and C, with nearly all of them started by the middle sib.[2] This series begins slowly. Just-hatched egret chicks jab each other lightly without causing any damage. At first they simply have no power. But within a few days, as the chicks grow increasingly nettlesome, nearly anything can provoke a skirmish. The most inflammatory thing a subordinate can do is raise its head above the height of its social superior, since pecks are usually delivered from the tallest position a chick can muster.

When one chick starts to get tall, a nestmate is likely to rise as well, the pair slowly extending upward like two flute-charmed cobras, keeping their bills horizontal and staring fixedly at each other, until they run out of neck. A moment of teetering bluff follows, a pause to see if the rival will back down, and then one throws the first punch, slinging its bill toward the other's face. The targeted chick leans back, like Ali in his prime, and may succeed in staying just out of reach as the attacking head passes by on the downward arc. As the first puncher draws its head back to its body and rises for another shot, the counter-punch often catches it midway. A flailing seesaw battle ensues, until one chick (usually the elder) makes solid contact on a few consecutive hits and the other concedes by crouching. After a week or so these fights can be quite protracted, and I have counted as many as 218 punches thrown in a single flurry. Even after crouching, a chick may suddenly regain its mettle, erect its head and initiate another bout. And then another. But eventually things get sorted out. The C-chick comes to avoid punishment by keeping its head down, and the fights gradually subside. Once the hierarchy is established, sporadic flare-ups are often settled by signal alone, a mere stretch upward being sufficient to send the subordinate back into a more seemly crouch.

Height is important for reasons other than fighting. Food also arrives from above. It is delivered in the form of boluses (five-centimeter balls of packed minnows or grasshoppers, depending on the egret species), regurgitated from the parent's storage stomach via pulses of reverse peristalsis, and advancing visibly up the serpentine neck. Every lump of food must pass from the parent's mouth, so the adult's bill is the prize that must be won. And because that bill is attached to a body many times taller than the competing nestlings, it might as well be lowered from a skyhook. Dramatically, the parent's head descends, and all the offspring know that this means food.

When the chicks are newly hatched, none of the fighting makes much of a difference. As noted, they cannot really hurt one another at first. And there is nothing really to win. Not only is the amount of food superabundant relative to the tiny chicks' appetites, but it is chunked out onto the nest floor where all can grab freely. Each neatly packed bolus is knocked apart by the first clumsy bill snatching at it and a scrambling melee follows, each chick downing as many rapid gulps as it can manage. The parent deposits several such prizes before the brood and waits until all three chicks are done, a state easily recognized by the shape of their necks. (A satiated chick is so crammed with food that its whole neck becomes inflated like an overstuffed duffel bag.) During the first week or so, the parent then reaches past its now-nodding offspring to recover whatever unconsumed bits of food remain on the nest floor.

The way food is delivered to offspring can affect the form that sibling rivalry takes. Here, great egret chicks receive about six modest boluses of regurgitated fish from a parent just back from the hunt. The A-chick has already swallowed the first two and dropped out of the fray, leaving B, C, and D to scissor the parent's bill in attempts to intercept the rest of the meal. The D-chick is missing feathers from its nape, where it has been pecked repeatedly by B and C. (Photo: Douglas Mock.)

After a few days, though, dining etiquette changes dramatically. Whereas even brand-new hatchlings show some interest in pecking and grasping at the parent's bill as it descends toward the nest floor, their attempts to grab hold of it with a strong, scissoring grip become increasingly spirited. Soon the A-chick masters a crucial skill: when the bulge of a bolus begins to roll upward, expanding the parent's throat, the A-chick seizes the parent's bill and uses that position to catch the food before it ever reaches the floor of the nest. This alters everything. From that moment forward, as each of A's nestmates in turn acquires the knack of intercepting food, more and more boluses are hijacked in mid-air. The competition ceases to be a scramble over food in the public domain (the floor) and becomes a contest for position, height, and thus control.

But if a chick has been losing a long string of fights, its options are meager. As the best-positioned chick scissors the parent's bill, initially out at the tip (which comes within reach first) then sliding up to grasp the all-important base, it improves its chance of controlling the food. The other nestmates often try grabbing the bill beneath the leader's spot, which helps them either catch something that spills during transfer or inherit the top grip if the incumbent slips off. Sometimes a detached chick tries to get above the leader, mouthing the parent's head over its eye and squeezing outward toward the mouth. The whole process looks like sandlot baseball players gripping their way up a bat to see who hits first, except that egrets do their best to cheat. And the B-chick may occupy itself pummeling C while A receives the first bolus or two. The number of boluses per meal varies greatly and is probably unknown to the chicks during a given meal, but four or five is pretty typical. The two senior siblings often get the first few; frequently they get them all.

The first week's fighting, then, is really groundwork leading toward the switch from indirect feeding (from parent to nest floor to nestlings) to direct feeding (interception), where the premium is on being intrepid. In fact, the earliest fighting carries no immediate prize at all and would be a waste of energy were it not for the subsequent transition to direct feeding. But intimidation pays off in the long run. Equally important, it does so in incremental units. That is, a nestmate need not be so thoroughly defeated as to cower in the corner to be out-competed. Each half-step that it hangs back, each moment that it hesitates, erodes its chance of getting the all-important grip at the parent's gape.

The switch from indirect to direct feeding is like the difference between how athletes gain control of a hockey puck during face-off versus a basketball during a rebound. The object of desire in a face-off is dropped to a specific spot on the ice and the two rivals can only swipe at it with their sticks. Swiping speed is the only variable. A basketball rebound, by contrast, typically caroms off the rim unpredictably, so there is more than speed, height, and timing involved. In the jostling beneath the goal, elbows have been known to fly as the players jockey for position: intimidation has to be factored in. (No doubt there is considerable intimidation in hockey, too, but it is hard to see how that can work during a face-off.) To finish the analogy with egret meals, imagine basketball with no rules against elbow assaults. You can imagine the sort of player who would triumph.

Egret sibling dynamics continue to change with each passing day, presumably depending on how things go with grocery deliveries. If the parents are inept providers and deliver fewer boluses, the food that reaches C (which often gets fed only after A and B have both eaten so much that they can no longer compete well, either in scissoring the parental bill or in hitting C) may decrease and even disappear altogether. When a shortfall persists for many days, C loses weight, strength, and much of its chance for survival. By the end of the second or third week, one-third to one-half of C-chicks have died; some others may be in jeopardy.

A few A and B chicks also drop by the wayside during the first few weeks, because of various maladies (such as gut parasites), accidents, and predation. Not counting broods where all the chicks perished (as can happen when something causes the parents to abandon the nest or when neighboring adults steal so many of the nest sticks that the whole set of eggs and young crashes to the ground), losses of individual senior siblings run at roughly 10 percent. When either senior sib dies young, C gets a promotion. Parents that might normally provide enough for two healthy seniors and a half-starved marginal chick can easily provide for C once it is part of a two-chick brood.

Thus the laying of that third egg carries at least two incentives for egret parents. About half the time it provides a fledgling, and some other C-chicks come through as replacements. The idea of laying an extra egg for insurance, which we encountered in the discussion of obligate brood reducers, can clearly be seen to apply also as a contributing incentive for overproduction in egrets and other species that practice facultative brood reduction. Indeed, one can partition the potential reproductive value that marginal eggs represent to their parents into these components: so much for the likelihood of surviving along with the full core brood, so much for the chance of getting through only after a core brood member fails.[3] One relatively unexplored aspect of insurance in facultative brood reducers is that the policy remains in effect longer than in broods of black eagles or white pelicans. Whereas eaglet victims last only a few days before succumbing, egret C-chicks that eventually die may persist for several weeks. At any point during that period, an opening in the core brood may occur, for example if a predator heists either of the senior siblings.

In Oklahoma cattle egret colonies, single chicks are often snatched by black-crowned night-herons, a distant relative within the taxonomic family (Ardeidae) to which the egrets themselves belong. As the name suggests, night-herons are more nocturnal than egrets, and for that they have very large red eyes with a well-developed reflective layer (the tapetum) behind the retina. This structure is a thin mirror that bounces light that has just passed through the retinal receptor cells (the rods, mainly, for black-and-white nocturnal vision) back through the same sensory cells again, thereby almost doubling the stimulation and hence the eye's ability to function in low light. The tapetum is what makes the eyes of cats, alligators, and many other nocturnal creatures seem to glow in the dark. And it enables night-herons to hunt well after twilight.

To me, it also gives them a malevolent look, especially as I watch one stalk toward a cattle egret brood I've been studying for days or weeks. Because I used to spend nights sleeping in the observation blinds (to minimize colony disturbance, among other reasons), I soon came to recognize the alarm cries of egret parents being menaced by an approaching night-heron. Rising from its brooding position, the egret faces the stouter heron, spreads its wings to shield the chicks, and yells, which is about all it can do. When the night-heron springs into the nest, the egret is usually bowled off its feet, and in the moment it takes the parent to dive back into the fracas, the grab is made and the night-heron vanishes into the darkness with a chick in its bill. The only winner, other than the predator, is the C-chick, unless it happened to be the one taken.

But unless a predator reduces their numbers, the nestling egrets generally resolve the supply-demand problem themselves. Rather little is known about the day-to-day behavioral decisions made by these players as the drama proceeds. But an interesting modeling exercise by Scott

Forbes and Ron Ydenberg, using a technique known as Stochastic Dynamic Programming, can help us think about the problems and suggests some testable predictions.[4] To get a feel for the approach, imagine that a senior sibling has to decide whether it should continue to tolerate the presence of its marginal nestmate (at least for now) or terminate that chick's existence. Imagine that the bully bases this choice on its best assessment of whether the parentally delivered resources are currently sufficient and likely to remain so down the road. The information available to the bully is bound to be imperfect, but logic suggests that its quality should improve with time. As the days roll by, first, the bully's experience with current conditions expands, and second, it has fewer days of uncertainty remaining before it will no longer have to depend on what the parents bring home. To take an extreme, if there are, say, fifty days in the nesting period and this is day forty-nine, then killing your still-viable nestmate is probably not ideal. But on day forty-eight that conclusion is a bit less obvious. Back on day thirty, the whole situation may be quite doubtful.

Way back at the beginning of the cycle, the first few deliveries provide considerable information. If food is inadequate way back then, when demand is as low as it will ever be, prospects for future abundance are decidedly bleak. Thus intimidating your nestmate as a first step toward eventual execution is probably more cost-effective than deferring the decision. But if food is plentiful at the outset, then tolerance may be prudent. Moving forward in time, then, each new day's food deliveries should allow an increasingly sharp picture of how prolific the parents really are (a process known as Bayesian updating). This kind of argument should also help us make sense out of obligate brood reduction and understand how complicated and variable facultative brood reduction must be.

There are compelling reasons for believing that food is the chief commodity over which egret nestlings fight and kill one another. Most compelling is the evidence of starvation. Victims tend to be emaciated, and even C-chicks that manage to survive grow more slowly than their elders. Second, though battling can occur at any time, it is nearly automatic when a parent returns to the nest with food. Through the first two or three weeks, parents take turns at hunting and brooding the young (or guarding the nest from a nearby perch), so one adult is present virtually all the time. Chicks may sporadically beg from the attending parent, but they do so less often as the hours roll by, because that parent is unlikely to comply (its stomach no doubt having been emptied). And the young are quick to spot the other parent, sometimes even recognizing it over impressive distance as it flies toward the colony. They sit up, then rise to stand on their toes, and wave their wings wildly, emitting loud triplets of unmusical squawks. And during this tumultuous greeting, B frequently starts to peck C.

If the dominance is well established, C's excitement over the imminent meal is quelled immediately and it crouches or rushes to the nest rim and hangs its head over the side (making that target harder to reach if B presses the attack). Other times, C may fight back and the two combatants may totally ignore the first bolus or two. Such altercations between B and C are good news for A, which wastes no time filling its neck. With C freshly vanquished, B turns frantically to scissor the parent. If C is completely intimidated, it may not even try to feed, but often it just hangs back a bit, conceding early boluses, before leaping in with a strong bid after both seniors have had something. One great egret C-chick I observed had a habit of scrambling completely out of the nest and hiding in a jungle of nearby foliage as soon as an arriving parent was spotted. There it waited, unscathed, until both seniors had ingested one bolus apiece, before hurling itself back into the center of things and competing fully, and often effectively, for the third bolus. As a former D-sib, I found much to admire in that bird's tactic.

Though always important, food abundance or scarcity is not the whole story on why egret siblings fight so hard. As noted earlier, the fact that this food can be effectively monopolized probably makes escalation cost-effective. During our second summer of watching great egret siblicide in Texas (1980), my assistants and I began to take notes on a second species, the great blue herons nesting in the same salt cedar bushes near the egrets. We had noticed that their chicks seldom fought and that their chicks were almost never bloody-naped, as egrets often were. Yet we also had the impression that the youngest and smallest chick often died anyway. By that summer's end, these impressions were well on their way toward being confirmed with data. During the first month after hatching, nearly as many great blue heron broods as great egret broods suffered chick losses,[5] but there were twenty fights in egret nests for every one in heron nests.[6] Why did the young herons not fight as viciously?

One conspicuous difference between the two species was that the great blue herons in that colony were living on a diet of large-bodied fish, the carcasses of which were presented singly to hungry broods. The average heron prey item was about 19 centimeters long and weighed about 200 times more than the mosquitofish and other small fry in the egret diet. Even with six to ten small prey items wadded together as a bolus, egret offerings arrived in tidy little packets easily caught in mid-air. Great blue heron offerings, by contrast, were seldom catchable. As the arriving parent moved onto the nest, heron chicks jumped up with full wing-waving and triplet-calling, but showed somewhat less enthusiasm for scissoring the parental bill. When the regurgitate emerged, it almost invariably slid to the nest floor, where the chicks jabbed at it. It seems that a large fish captured by a heron is swallowed headfirst (this precaution facilitates passage of the stiffer fins while the dying fish might still try to erect them), but settles only partway into the stomach's acid-bath. Chemical digestion works initially on the head, while the tail remains relatively firm and unaltered.

Heron chicks quickly learn all about this asymmetry in the topography of fish edibility, adjusting their aim from a random early scatter of pecks to a keen interest in the head. Within a few days it takes only a minute or two of rapid chipping to remove all accessible chunks and gobble them down. The parent then swallows the rest of the carcass again and cooks it for another hour or so before bringing it back up for another round of chipping and gobbling. Until its mate relieves it, the same parent (and the same, slowly shrinking fish) continues to feed the brood intermittently.

We have no reliable data on how much individual heron chicks ingest (these nestlings are harder to see than egret chicks when very young because of higher nest walls, and their chunks of fish tissue are harder to quantify than the discrete egret boluses), but there is obviously not a lot of useable food to go around. In this heron population, then, brood reduction seems to result from a scramble competition resulting in eventual starvation of the runt. The senior siblings are more adept at carving off enough fish tissue to support proper development, while the youngest sib often becomes emaciated and eventually wastes away.

We tentatively attributed the difference in aggressiveness between egrets and herons to the differences in prey size for the two species. For egret chicks intimidation is feasible because the food boluses are small enough to catch in midair; heron chicks are more like hockey players in a face-off. This explanation was tested in two ways. First, I performed a cross-fostering experiment in the Texas colony, swapping whole broods of just-hatched great egrets for their great blue heron counterparts.[7] A total of ten heron broods thus received their rations from egret adults (getting tiny prey in discrete boluses), while ten egret broods got to face those big fish with the predigested heads. All parents accepted their fuzzy broods readily.

Great blue heron chicks raised by egrets performed as if they had read the egret script. They quickly became obsessed with scissoring their surrogate parents' bills and soon mastered the art of direct feeding. They also became bellicose with one another, fighting at rates very close to the egret standard (and fifteen times more often than herons normally fight). The usual proportion of C-chicks died, but this time bearing telltale nape wounds. For egret broods raised in heron nests, though, things were much less clear. These tiny chicks persisted in trying to receive prey using the bill-to-bill direct method, despite the fact that the prey items were many times larger than the chicks trying to catch them. The egrets' behavioral rigidity had unfortunate consequences as the nestlings sometimes missed their chance to peck off edible chunks for the miserable reason of being pinned beneath the fish carcass while nestmates fed. Many of the egrets became caked with fish slime and most found little to eat. The mortality rate of C-chicks was within the usual range, but their fighting dropped only about 20 percent, too small an effect to be statistically demonstrated in this modest sample.

The next step in the plan was to balance the artificial nature of the cross-fostering experiment with a pair of field studies. One study would explore how great blue heron nestlings behave in parts of their range where heron parents happen to provide monopolizably small prey. The other would investigate egret brood dynamics in parts of the species range where parents naturally deal with large prey. I had some evidence that the herons of northern Quebec eat tiny sticklebacks, while the egrets on Isla Tabóga, Panama, were said to prey on large fish. If so, then I hoped to see if the herons' behavioral flexibility observed in the Texas experiment might be more evenly mirrored by egret broods in regions where that species was accustomed to a large-prey diet.

As usual, these studies provided mixed results. The trip to Panama lasted just three days. In that time I learned: (1) that my information on diet size was in error (the Panamanian egrets were eating small fish, as in Texas); (2) that the western Pacific Ocean was in the throes of an exceptionally strong El Niño warm water event, which meant (3) that the last few remaining breeding pairs were in the process of abandoning their freshly hatched broods. I barely had time to wave good-bye as the parents departed.

The June 1984 trip to Quebec, by contrast, ran like clockwork. Rick Williams, my field assistant and traveling companion, and I drove my twenty-year-old pickup truck 2,071 miles from Oklahoma to the tiny French-speaking village of Forestville, towing the research boat. I had chosen the local heronry, not far offshore in the vast St. Lawrence River, for one outstanding feature. Though the nests were 20–25 meters up in mature pines, the small island's center consisted of a massive granite dome. We made a trail up the back side of the dome and could look down into the nests from a 2–5 meter height advantage. Rigging some army-surplus tank camouflage netting between two trees, we fashioned a blind that we could enter from the woods without the birds' knowledge to minimize disturbance. Still, we had to clear all branches from the lines of sight between our blind's position and the seventeen nests available for study. And, as usual, we had to color-mark the hatchlings with small patches of dye so we could keep track of each chick's rank. These tasks involved climbing the trees using leg-spikes, something I had done previously in lesser trees, but these pines were too thick at the base for my hug-the-trunk method. My choice of Rick was not accidental, as he was a seasoned rock-climber and had brought along all manner of ropes, caribiners, harnesses, and so on for just this purpose. Soon we were moving methodically up and down the trees, getting ready for the hatchlings to make their appearance.

From our perch atop the granite dome, we watched as the drama unfolded pretty much as expected. The fish regurgitated for hatchlings were mostly sticklebacks small enough (less than 10 centimeters in length) to allow interceptions. So the heron chicks performed the full scissoring routine, learned to catch boluses, and made the complete transition from indirect to direct feeds, as the egrets and egret-fostered herons had done in Texas. And fight they did, an average of 8½ times more frequently as the Texas herons with the large-prey diet.[8] In short, the link between small, monopolizable prey and the effective use of aggression looked pretty solid.

It is interesting to contemplate why the great blue heron chicks were able to turn sibling aggression on and off in response to prey size, while great egret chicks appeared less flexible. The simplest reason I can think of is that the diet of this heron species may be more variable than that of this egret species. The difference between the Texas and Quebec heron diets may occur from time to time more locally (for example, if large fish are in short supply in the shallows during some Texas summers, forcing Texas herons to capture larger numbers of small fish). Also, some great blue herons have been reported to specialize as upland mouse-catchers: do the offspring of such parents use aggression to advantage? The flip side of that explanation is that it rests on the assumption that great egrets seldom if ever rely on prey species too large for their chicks to monopolize by scissoring. Thus, if anyone were to locate a population of great egrets that routinely capture oversized prey, the pugnacity of their nestlings would be well worth documenting.

Setting aside the issues of prey size and monopolizability, a key aspect of any competition over food is the sheer amount provided: the less there is of some limiting resource, the more acute the rivalry. As noted, there is good reason to believe that food is often the precious limited commodity. So food shortage or risk thereof seems likely to be the driving evolutionary force behind offspring selfishness—the ultimate cause of all the aggression, intimidation, and execution. The specter of starvation hangs over the whole drama. From that realization, it is a short mental leap to imagine that food shortages might also provide the *immediate* stimulus, the proximate cue for the fighting. The argument works out as a proposed simple behavioral rule that goes something like this: "If hungry, then fight; if satiated, relax!"

This commonsense expectation was simply assumed by most biologists, myself included. David Lack even went so far as to state this as fact (as if he or anyone else had actually bothered to check): "The predisposing cause of death is food shortage among the nestlings, since *a well-fed chick does not attack its nestmate.*"[9] End of issue. The beauty of the supposition is that even the bully comes out ahead, redirecting its energy into growth and maturation when sustained fighting is probably unnecessary.

The pile of hyraxes in that black eagle nest probably should have warned us that the matter might be more complex, but it did not. Nobody was thinking much about preemptive strikes or the merits of hierarchy maintenance.

Some pioneering field research on facultative siblicide in kittiwakes, an odd gull that nests on narrow cliff ledges, delivered data that looked like a good general fit to Lack's reasoning.[10] In their Aleutian study colonies, Barbara Braun and George Hunt found that the precipitous nest site gave the A-chick the option of pecking and shoving B off the ledge, such that gravity and sharp rocks below complete the siblicide. More B-chicks vanished during periods of foul weather (when parents presumably delivered less food to the brood) than when conditions were more propitious. It was assumed, but not verified directly, that the key proximate effect of the low-food periods was increased aggressive pushing by the A-chicks. A similar pattern of victim sibs dying during inclement periods had also been reported for ospreys, a fish-eating hawk, and again the hunger-mediated aggression was assumed initially.[11]

But something seemed out of place with this worldview when one contemplated great egrets. Six captive broods that we hand-fed beneath our stilted beach-house in 1979, for example, were totally satiated three times a day, yet they continued to fight. Because we weighed each of those chicks before and after every meal, we knew how much they were eating, and this turned out to be six times more food than they get from real egret parents. In such luxurious conditions, why were they not relaxing?

After a few summers of observing great egrets, it was obvious that some parents were a lot better at catching fish than others, so that some broods were well fed while others were not. I used those records to seek a possible negative correlation between total food delivered and fighting rates, simply checking whether chicks in high-food families tend to be less aggressive than those in low-food ones. Lo and behold, the ones that ate more appeared to fight a bit *more,* not less. It began to seem as if the egret fighting rule might be more like "If hungry, save your precious strength; if well fed, use some of that surplus energy to assert control." This was so counterintuitive that I later ran similar correlations on other data sets (for example, the seventeen Quebec great blue herons broods and cattle egret data sets from Texas and Oklahoma). Each time, I found weakly *positive* relationships between food level and aggression.

This emerging pattern led to a field experiment involving 100 nests in the Texas colonies one summer. At half the nests, Tim Lamey and a field crew doubled the broods' daily diets by supplementing them once a day with a jumbo-sized meal. As the crew approached each experimental nest, they placed a pre-measured wad of food in one end of a 1.5-meter length of plastic rain-guttering (small and chopped fish obtained from the local shrimpers, equal in mass to the amount two average parents deliver over the whole day). They extended this low-tech apparatus over the nest rim and then inverted it, allowing the lump of food to fall onto the nest floor. The chicks hastily grabbed the food. At a second, randomly selected sample of 50 control nests, the ceremony with the same gutter was repeated *sans* food. A subsample of the nests, 13 experimental and 10 control, was close enough to the blind that detailed behavioral records could be made. From these records we confirmed that the intended consumers were, in fact, eating all the food. We also documented that the parents did not simply cut back on the amount they were providing (which had been a worry), but continued to bring the same level of their own deliveries as before.

Our artificial doubling of food intake had dramatic effects on the growth and survival of C-chicks. By their tenth day, C-chicks in the provisioned broods were more than 60 percent heavier than their unprovisioned counterparts, and their survival through the first month was significantly improved. But sib-fighting did not change significantly. It thus appears that great egret nestlings really do not predicate their attack behavior on current food levels, at least not in the expected direction: bullying continues unabated in times of plenty. What does change is the designated victim's future, which is considerably brighter when food is abundant. Not unreasonably, the C-chick is better able to withstand the inevitable beatings when it is getting enough to eat. In short, aggression as such may not be the key facet governed by current food levels.

Exactly why senior sibs do not reduce their aggression when they have plenty to eat is an interesting secondary puzzle. To me the most plausible explanation, which is far from demonstrated, involves tailoring the argument about pending competition to fit facultatively siblicidal species like egrets. If the A-chicks of black eagles and white pelicans kill automatically and quickly (preemptive strikes because food is exceedingly likely to run short), then perhaps egret bullies use sublethal doses of aggression for a parallel reason. If they cannot know whether enough food will be supplied in the future, elder sibs have reason to make sure the intra-brood hierarchy is in place and well understood. And they do this early, when it presumably is easier and cheaper to achieve. Birds in warlike families do learn their roles, with an individual's recent victories likely to make it more aggressive; recent losses, less so.[12]

As a preliminary test of this line of thinking, I worked through the data on great egret food delivery to see whether deliveries are predictable over the short term. I performed a statistical exercise called a runs test that can detect whether a given day's amount of food at each nest was at all predictive of that same nest's deliveries three days later. It was not. So the egret system may work by senior chicks establishing themselves atop the pecking order at the very beginning, in case being dominant may make the difference between survival and death.

There are, however, other species that do seem to adjust their aggressiveness in response to current or recent food levels, just as Lack expected. In two-chick broods of blue-footed boobies off the coast of

Mexico, Hugh Drummond and his students discovered that the A-chick escalates its normally moderate rates of pecking B when its own growth becomes undesirably slow. Specifically, when A's mass drops to about 75–80 percent of what it could be at a given age (the maximum achieved by same-age A-chicks), the B-chick tends to die, regardless of whether B itself is underweight.[13] Of course, such a mortality hike might be due to starvation (less food trickling down to the B-chicks) and not necessarily to heightened abuse, but this time a compelling experiment brought in a guilty verdict on the A-chick.

Whereas in our experiment with egrets we sought to turn fighting off by adding extra food, the question for the boobies was whether aggression might be switched on when food became rare. Hugh and his student Cecilia García temporarily reduced the food intake of a sample of A-chicks by wrapping cloth tape around the chicks' necks. This unyielding tape was not tight enough to cause discomfort in breathing or to interfere with any aspect of behavior other than swallowing. Because the fish regurgitated by parents consists of food *lumps,* the chicks' necks must expand laterally when they eat. For the control broods, chicks were similarly taped, but the tapes were removed for meals (remember that Drummond studies very tame island-living birds). In short, then, experimental A-chicks were put on a crash diet while control A-chicks were allowed to feed, and both were watched very closely. After the first day, food-deprived A-chicks lost weight down into the 75–80 percent zone and their sib-pecking rate quadrupled, but the pecking of control chicks remained unchanged.[14] This was the first time any bird had been clearly shown to escalate aggression in response to food deprivation.

A second siblicidal bird has also been shown to do so. As mentioned earlier, osprey victim chicks tend to succumb in bad weather. But provision of extra food has been shown to reduce the bullies' pugnacity. Working in eastern British Columbia, Marlene Machmer and Ron Ydenberg devised a clever way of getting the necessary data. Their problem was that ospreys commonly nest in tall, well-scattered, and frequently dead trees that are hard for climbers to reach, let alone observe. But two "viewing nests" happened to be atop pilings in the Kootenay River within 100 meters of a high and sturdy railroad bridge; ten others were relatively accessible by boat. Osprey parents were obligingly tolerant of having their broods exchanged for two-day experimental tests, during which each brood was sampled up to six times. These test trials were preceded by focal broods being isolated for 3.5 hours in an artificial nest (to allow the chicks' stomachs to empty) then either treated to a full hand-fed meal of small fish pieces or sham-fed by equivalent hand motions that carried no food. The trials themselves were done in the viewing nests, where the focal broods were deposited along with a fresh fish. The resident female quickly returned after the disturbance of the boat's visit and performed the ritual of tearing off bite-sized pieces of fish for the chicks. The observers on the trestle counted how many pieces of fish each chick got, and in what order, plus all acts of aggression among brood members. Because each brood was used several times, with alternating "fed" and "sham-fed" treatments, each served as its own control. The results were quite clear: excluding one brood that could be tested only once and another that never showed any aggression, seven of eight broods were more aggressive after sham-feeding sessions than after meals. Because this pattern is unlikely to have occurred by chance alone, it strongly supports the view that hunger is linked to aggressiveness.[15]

Short-term sensitivity of sibling aggression to hunger has also been demonstrated in black guillemots, small burrow-nesting seabirds. By placing small video cameras inside the nest chambers, field workers could observe the interactions of 15 two-chick broods that initially had regular parental visits and food deliveries (for three hours), but then experienced a six-hour hiatus with no food (parents were frightened from burrow entrances by balloons with painted eyes and/or a fiberglass predatory gull), before parents were allowed back in (for three hours). While the parents were excluded, A-chicks increased their attack rates moderately (from about one attack per hour to two), but that went up steeply (to more than seven attacks per hour) after parents were allowed to resume feeding and there was something to contest. After their six hours of deprivation, A-chicks consumed about 75 percent of the food, as opposed to roughly 50 percent in the pretest and once things had settled down the next day.[16]

To summarize, in the only four siblicidal birds for which the proximate link between chick aggressiveness and hunger has been tested experimentally, it has been demonstrated nicely in three (blue-footed booby, osprey, and black guillemot), but seems lacking in a third (great egret). While I was surprised at first not to find it in great egrets also, the quantity and quality of evidence in that species has led me to the view that siblicidal species differ fundamentally in this respect.[17]

If great egrets really do not use hunger as their cue for adjusting pugnacity (perhaps because current food amount is not a reliable harbinger of future conditions), we must wonder if these chicks are condemned to battle endlessly. It seemed that they should have some way to turn off their aggression when their family is in stable and beneficent circumstances. What might signal them reliably that a truce is safe? For this answer, we turned away from the food base entirely (the supply side of the dynamic) and began to scrutinize demand. We had a clue guiding us in this respect. From our very earliest observations of great egrets it seemed that just after the death of a victim chick the two surviving sibs tended not to fight each other much. This could happen, we reasoned, either because egrets had some crude form of counting or because the dominance relationship between A and B was already clear and did not require maintenance.

Tim Lamey and I decided to test these possibilities with cattle egret broods in Oklahoma (which had shown comparable calming after the death of a victim chick) using a clever experimental design proposed by my wife and colleague, Trish Schwagmeyer. The idea was pretty simple: first we would obtain a baseline picture of how hard chicks in each three-chick brood were fighting normally, then "eliminate" one (as if a predator had grabbed it or it had died in an accident) and see how hard the two "survivors" continued to fight. Finally, we would restore the kidnapped chick to the nest in the pink of good health, and see if the hostilities renewed. We did this for ten experimental broods in three-day phases. First we recorded baseline activities for three days before visiting the experimental nests and pocketing the still-healthy C-chicks. We carried these C-chicks to the lab and fed them a diet of lab mouse carcasses through that phase, before restoring them to their home nests for three final days of observation. Thus each brood served as its own control: if brood size in itself is an important cue that reduces fighting, then per capita fighting during the two-chick phase ought to be sharply lower than in either of the other (three-chick) phases.

What made Trish's design truly elegant was a second set of control nests, a luxury affordable only because cattle egrets are so numerous in colonies that we had many broods to spare. During the full nine-day study cycle, we simultaneously recorded the food-and-fight dynamics at ten more nests where brood size was only sham-manipulated. That is, after the three-day baseline period, we visited these ten nests, briefly removed the C-chick, kissed it on the tip of its bill for good luck (instead of pocketing it), and put it right back where it had been. These control broods were similarly visited and briefly handled when the experimental broods were being restored to full size.

The effects were even more dramatic than we had imagined. When broods were trimmed to two chicks, fighting all but vanished (an average decrease of 95 percent), but it rebounded right back to control levels when the C-chicks returned. Fortunately, Tim was thinking more clearly than I during his shifts in the observation blind, and toward the end of the study he called a major problem to my attention. Might not the *rank* of the removed chick be important in producing the cease-fire we observed? Specifically, since we had stolen only C-chicks (which we knew were the preferred targets for most of the attacks by the pugnacious B-chicks), we might have taken away the one player wearing a bull's-eye. This was such an irritatingly likely possibility that we simply had to repeat the whole experiment with a whole new set of nests, spending another nine hot days in the blind, but removing the A-chicks instead. This left the most combative dyad intact. Again, we ran a second set of sham-removed controls. And once again, fighting plummeted during the reduction phase and resumed after A's return. (Actually aggression in the experimental broods became a bit inflated over control levels, as A reasserted itself over the B-chick that had assumed the top slot.)[18]

Thus it appears that cattle egrets quickly curtail battling when brood size drops and, presumably, demand for food is reduced to a level that parents are able and willing to meet. This number-of-chicks mechanism may seem cruder than sensitivity to hunger as a means for siblicidal birds to adjust the balance between supply and demand, but it is also a lot more reliable in the sense that the disappearance of a nestmate is (ordinarily) permanent and irreversible.

The parent egrets did one other thing during this experiment that caught us off guard, but it made interpretation of the proximate trigger for sib-fighting a lot easier. When brood size dropped from three to two, in both trials, parents neatly cut back on the amount of food delivered so that the per capita volumes of grasshopper boluses remained unchanged. This means that the chicks' cessation of fighting could not have been a response to bonus food as such (because they got none); it had to be due to diminished numbers of nestmates. But as usual, when one mystery emulsifies, another gels. The riddle switches from the behavioral mechanisms that control aggression to the evolutionary reason for such to exist. The two chicks remaining in the nest during the removal phases did not get food for three, as we had expected, but an amount of food adjusted by the parents for the smaller number of hungry mouths. So what does this suggest as the fitness incentive for siblicide in the first place?

That is, what's in it for the bullies? At this stage, we can only speculate, but the logic follows thus: First, if parents normally stretch themselves a bit thin trying to support a marginal chick that is likely to be lost eventually, its demise relaxes strain on the family budget. If parents pocket that portion of the food, digesting it for their own body tissues, they are more likely to breed again in the future. When they do, these two surviving core chicks acquire siblings that carry identical copies of their genes. In short, parents have the potential to contribute to the victorious chicks' inclusive fitness. More immediately, parents that *can* deliver more food than they choose to share (like those in our study) are also better bets to get the two survivors through temporary hard times in the weeks ahead (for example, during inclement spells). The core brood is more safely buffered, affordable through a broader array of ecological ups and downs.

We have seen that uncertainties about food, dictated by environmental variations on a broad scale and adjusted also by some parental decisions, have differing effects on the way sibling aggression is used in different siblicidal birds. In some species the fighting itself is reduced or shut off in times of plenty, while in others fighting remains relatively steady until demand drops. Next we shall meet some birds that compete on a scramble basis, without overt fighting at all.

9

Gambling with Children

Proverbial wisdom counsels against risk and change. But sitting ducks fare worst of all.

—Mason Cooley

Across their immense North American range, red-winged blackbirds are in the business of converting the tissues of aquatic insect nymphs—plucked bodily as they first emerge from the surface waters of marshes, ponds, and sloughs—into baby blackbirds. This is quite a famous bird, both for beginning birdwatchers who find the brightly plumaged males mercifully easy to recognize and remember (what *else* could you call a bird that is jet-black except for blood-red epaulets?), and for researchers who have focused much effort on trying to understand why these males are so much showier than their sparrow-drab mates.[1] The study of avian mating systems is indeed rich in detail about how redwing adults get it on; in the parlance of science, this is a model system.

The redwing is the textbook example of "resource-defense polygyny," a mating system in which each male strives to control key limiting resources that females need for breeding, thereby attracting multiple females. Last time I checked, the redwing record for simultaneous female partners stood at seventeen.[2] A male's ability to corral resources depends on features of the environment, especially the way key resources are distributed in time and space. When food and suitable nesting sites exist in dense pockets then females are likely to be similarly clustered.[3] With some males having many mates, there are going to be lots of have-not males, guys left out in the cold so to speak.

As with all species that are both sexual and genetically diploid, each and every redwing offspring has precisely one mother and one father. If the adult population is composed equally of males and females, then polygynous mating must have very different impacts on male and female fitness scores. In general, it is a buyer's market for females, virtually all of which are guaranteed to find a mate. The main problem for a given female redwing is deciding *which* male, from a marshful of courtiers, offers the most attractive total incentives package.

For males, the problem is obviously reversed. While essentially all female redwings have the opportunity to breed each season, the male side of the same tally sheet is filled mainly with *zeros* (many, probably most, males fail to get territories at all, and nearly a quarter of those that do produce no fledglings for various reasons), a bunch of *ones* (for males able to secure a single partner), and some much higher scores for the successful polygynists (in the Columbia National Wildlife Refuge 3–4 percent of males had ten or more females).[4] When all the male successes are tallied up, their total has to equal the sum of all female successes. It follows that if the adult population is balanced between males and females, the overall male average or mean success must also match the female mean. But the measure of statistical scatter between the two sexes, the "variance," is much greater on the male side of the ledger, because of all those zeros and the occasional super-stud scores.

In this heightened and asymmetrical variance lies the key to understanding what Darwin called sexual selection. The nutshell explanation for why males are bright and showy while females are drab, then, is that the process of mate-acquisition acts much more forcefully on these males than on their mates. In particular, any male feature that has the effect of making its owner a sexual have-not is likely to have its genes decrease in the gene pool. Meanwhile, features promoting above-average mating scores are likely to increase in frequency. And, of course, alleles underlying the traits that lead to the few super-high mating scores are replicated like crazy. There is considerable debate on precisely how these differences lead to remarkably exaggerated traits, but little doubt that this mating variance lies at the heart of the puzzle.

For adult redwings, in particular, it appears that sexual selection has led to the male acquisition of very bright shoulders, loud voices, and nasty attitudes toward one another, while laying a more relaxed hand on the plumage of females.[5] By contrast, *natural* selection, the broader process that shapes traits to enhance survival, chick-rearing, and everything else beyond mate-acquisition, seems to have molded the appearance of female redwings with more of an eye to camouflage during the vulnerable periods of incubation and chick-feeding. Their mottled brown and tan plumage blends easily with dappled shadows cast by the bulrushes and cattails that arch over their open nests, giving them a measure of local obscurity even in the middle of a dense colony.

This, at least, was the story on sexual selection in redwings for quite a while. Getting another female or two to nest within his territory was viewed as the male's chief preoccupation. Once on board, she was confidently tallied (by biologists, that is) as one more unit of the male's success.

Of course, it turns out that the females are not mere chattel of the resident potentate. In a classic bit of research serendipity, it was discovered that females do not restrict their sexual favors exclusively to territory owners, as tacitly assumed. In the early 1970s, the fact that this blackbird is an important grain pest in its winter range inspired an unsuccessful attempt to curb its populations via surgical vasectomies. The reasoning was that sterilizing the males holding the largest harems would produce a great many infertile eggs and reduce the whole population. The failure of these directed vasectomies to block fertilization opened our eyes to the existence of blackbird cuckoldry: clearly, females in the sterilized males' territories were obtaining viable sperm from other males.[6] More recent work on this point, using DNA fingerprinting, has revealed a maze of mixed paternity within and between broods, with no clear pattern as to which males come out ahead genetically.[7]

But enough of this discussion of sex, which is tangential to our topic except in two key respects. First, if two offspring in a redwing nest are just maternal half-siblings (same mom, different dads) rather than full siblings, the relatedness between them, *r*, drops from .50 to .25, making each half as valuable to the other for inclusive fitness. Even if the chicks cannot discriminate between full and half-sibs, as some mammals can do, the impartial scorekeeping of natural selection would be based on the weighted average. For redwings in three DNA studies,[8] for example, roughly one-quarter of all nestlings were half-sibs, so mean relatedness among nestlings in those populations drops only modestly, to roughly .44.[9] Perhaps more important, in most redwing populations all the incubation and most, sometimes all, of the food for the nestlings is provided

Four very young red-winged blackbirds open their mouths and stretch upward as their mother approaches the nest. Sibling rivalry in many songbirds seems to consist mainly of such postures and accompanying vocalizations, with most of the decisions about which chick to feed being made by the adult. (Photo: Scott Forbes.)

by the mother working alone.[10] Being raised by just one food provider can pose supply problems for the brood, and we are always alert for signs of tightness in the family budget that might affect social interactions.

Like most other songbirds, redwings hatch their (usually four) eggs in a semi-asynchronous pattern, which means that one chick is a day younger and smaller than the others. Death comes to baby blackbirds in many forms, ranging from violent hailstorms that can smash all the nests and eggs (and sometimes the parents as well) to hungry water-snakes that glide up the cluster of cattail stems supporting the nest and swallow all non-flying inhabitants. Mortality also results from food shortages and starvation. Broods that manage to get past the forces of mass execution qualify for a possibly fatal round of sibling competition.

Suspecting that semi-asynchronous hatching may serve the same functions for redwings that longer hatching intervals serve for other species, Scott Forbes and his students at the University of Winnipeg have been marking individual redwing eggs, chicks, and parent redwings for several years. Forbes postulated that the late-hatching egg might be a project that parents take without necessarily being able to complete. As in the larger birds we have already encountered, this family member appears to be marginal, the others representing the more affordable core brood. To put it the other way around, perhaps the parents handicap the last egg to allow it to be discarded more easily.

Redwing nestmates do not hammer and cannibalize one another, nor do they serve as helpers for each other later on. A marginal chick may provide some ephemeral (and reciprocated) thermal advantages by reducing the surface-to-volume ratio of all nestlings, since their physiological ability to regulate body temperature only becomes fully functional well into the fourteen-day nestling period. But the potential for helping others is unlikely to provide the primary reward to parents for aiming a bit high and creating that extra egg in the first place. The key incentive probably lies more within the realm of insurance value or resource tracking, or more probably a mix of the two.

As with other colonial birds, it is relatively easy to tag redwing nests and build large data sets. Scott's group first assessed the insurance value of marginal eggs by elaborating on the experimental approach that Kevin Cash and Roger Evans had pioneered with siblicidal pelicans, adapting it to a facultative brood-reduction system. If the marginal egg serves as insurance, then its life chances ought to improve sharply when a core sibling is experimentally removed, but the reverse ought not to be the case. That is, the head start enjoyed by core offspring ought to guarantee their doing well, regardless of a lesser nestmate's fate. But a marginal offspring's situation should be far more sensitive to what happens to core siblings. Of course, if the marginal egg does *not* carry insurance value, no such asymmetries are expected.

In the first part of the study, it was important to establish that the initial prospects for core and marginal nestlings are very different. Excluding cases where the whole brood was lost simultaneously (to predation, storms, and so on), a very large sample of unmanipulated (and sham-manipulated) broods revealed the basic caste system. Fewer than 7 percent of core offspring died during the first week of life, while more than 35 percent of the marginal young were perishing.[11]

The next step was to see how inserting or eliminating nestmates would affect players in these roles. The high success rate for core chicks was neither threatened by the experimental addition of an extra marginal chick (core mortality rate rose by a trivial .7 percent) nor relieved by the removal of a marginal chick (core mortality eased down by a statistically meaningless 3.1 percent). Even adding an extra core nestling did not affect mortality for the other core chicks enough to matter statistically (6.3 percent increase), and removing a core nestmate was also negligible (core mortality down 1.7 percent). Thus, from the perspective of core brood members, life is exceedingly steady.

Not so for marginal chicks. While adding or subtracting a single marginal chick did not affect the survival rate of a remaining marginal chick, adding an extra big kid raised the likelihood of a small kid's dying. Most impressively, and apropos of the insurance hypothesis, removing a single core chick from each brood led to a 31 percent jump in the likelihood that a marginal brood member would survive. From the mother's point of view, then, it seems clear that an extra egg pays a handsome dividend in terms of protecting her from raising an undersized family (smaller than what she can typically afford). Semi-asynchronous hatching establishes a modestly handicapped designated victim that is sometimes unable to keep up in the begging competition, but if any of the three core chicks happens to die, that second-stringer is ushered in from the periphery to occupy one of the likely survivors' slots.

As we would expect, then, the payoff for parents of facultative brood-reducing species is not so limited as for their obligate counterparts. Unlike the situation in black eagle families, the death warrant for an extra offspring in a redwing nest is conditional, a gamble apparently set to err on the side of optimism. If there are often enough resources for all four blackbird nestlings to fledge, the pressure for decisive, early resolution of temporary or pending competition is relaxed, presumably outweighed by the potential for higher gains in inclusive fitness if all can thrive. With extra days for family members to assess the ups and downs of parental food deliveries, the marginal chick that began life primarily as an insurance policy may turn out to be a truly affordable luxury without threatening members of the core brood.

If one were to ask field biologists the question "Why do parents commonly overproduce?" the great majority would be unlikely to mention insurance at all. Instead, they would point to the chance that all offspring might survive, the resource-tracking incentive. This illustrates the overwhelming influence of David Lack's highly original 1947 paper. Resource tracking has been the standard and, indeed, pretty much the stand-alone explanation for over half a century. I opened this book with an overview of multiple hypotheses and a discussion of insurance (by focusing first on obligate brood-reducers, for which Lack's idea cannot work) in part to rectify this historical bias. And I am greatly amused, now that nonscientists are beginning to show interest in this area, to see that most newcomers have no difficulty accepting the insurance argument. That is, without knowing what the established orthodoxy says, what the answer is *supposed to be,* novices quickly embrace a pluralistic view of the problem that formally trained biologists, myself included, have understood only fairly recently.[12]

In hindsight, it can be seen that Lack provided a subtle and deeply satisfying ecological answer, based on unpredictable food levels, to the issue of overproduction. More to the point, he did so before anyone else realized there was a question.[13] This was characteristic of David Lack in several respects. First, he was way ahead of his time on several fronts. Second, he was inclined to find a food-based explanation in an era when many of his colleagues were infatuated with predation pressure. Working on common swifts, small acrobatic insectivorous birds that conveniently built their hard mud nests in the Oxford bell tower, Lack showed that parents could often rear a greater number of truly healthy young when they have two chicks than when they have three. This is called a "density-dependent" effect, meaning that there is a limit on some key resource (such as food) that inevitably creates a trade-off between offspring quality and quantity (density in the nest).

A great debate during Lack's heyday concerned how whole populations of animals and plants are held in check instead of growing to infinity (recall Darwin's millions of elephants from the epigraph to Chapter 1). Two camps polarized (as rival camps are wont to do) between ecologists espousing the superior power of density-dependent forces (read: competition for resources) and those more impressed by density-independent ones, such as protracted but unpredictable famines and pestilence that sporadically sweep away most members of a population. This controversy came clearly into focus in 1954, when two remarkable opposing books were published, Lack's *The Natural Regulation of Animal Numbers* and one by a pair of Australians, H. G. Andrewartha and L. C. Birch, entitled *The Distribution and Abundance of Animals.*

As entomologists, Andrewartha and Birch naturally focused on insects, which they regarded as seldom checked by food supplies, but more by the end of each summer's easy weather. They pointed out that insect populations build up rapidly during the breeding season, then are knocked back severely by winter. The slate is wiped clean. For example, a small insect named *Thrips imaginis,* which feeds on crop plants in the temperate climate of southern Australia, does very well during the plants' breeding season (it eats pollen). Thrips populations rise sharply, but when plant breeding conditions deteriorate, the insect densities drop as well. During the winter many of the slow-developing thrips fall to the ground inside the withering flowers they inhabit and die in massive numbers. So it is the ebb and flow of seasonal change, especially of warm weather, that lifts and drops the insect.[14] Moreover, even at their peak thrips do not seem to keep up with the crop plants. The warm season triggers an exponential growth in their populations but does not last long enough for them ever to hit the resource ceiling (the so-called carrying capacity), indeed they barely make a dent in the total food supply available to them.[15]

Meanwhile, half a world away in a moister and relatively stable habitat, David Lack was studying parental care in birds that go to great lengths to buffer their offspring from environmental vicissitudes. Birds do not die off sharply each year, but work close to the carrying capacity by raising a few new offspring and investing heavily in each.

We now can accommodate both density-dependent and density-independent forces in nature and recognize that they affect different life forms quite differently. In fact, with the great advantage of a half-century's distance, the whole argument seems to have about as much intellectual meat as the inane beer commercials in which opposing gangs of chunky ex-jocks shout back and forth over whether the product's greater merit is "Tastes Great!" or "Less Filling!" In fact, even the protagonists of the mid-1950s debate over density effects were less divided than they thought themselves to be. The Australian entomologists knew full well that the thrips populations dwindle during the increasingly adverse period at the end of the spring bloom, when the insects are gradually squeezed into the most favorable bits of microhabitat, an annual period of density-dependent bottlenecks.[16]

In the organisms we're calling nursery species, the dominant force within the nursery is clearly on the density-dependent end of the continuum. There may be important parts of the parental fitness calculation that have to take into account whether the nursery itself is going to be smashed by hail or plundered by predators, but that is on rather a different scale. For the most part, density-independent catastrophes seem less likely to shape intra-family social relations directly.

So, to review, Lack's idea was that parents create an extra offspring (or two) in order to capitalize on the unknowable chance that food may turn out to be sufficient for all, but they build in a pressure-release valve by handicapping certain designated victims, the "marginal" sibs. Unpredictably fluctuating resources are seen as driving this parental strategy. For example, though two may be the largest number of chicks a pair of common swifts can raise properly in an *average* season, the occasional windfall can make three affordable. And if that situation arises often enough, the habit of laying three eggs may be supported by natural selection.[17]

Lack drew attention to the ubiquitous trade-off between offspring quantity and offspring quality, focusing on the latter without forsaking the former. His more general idea in this area was that natural selection favors parents reproducing as rapidly as possible[18]—a view completely at odds with the pervasive assumption of his time that parents show reproductive restraint in order to preserve their species, to protect it from possible extinction that could follow if everyone tried to maximize breeding. Contrary to those who contended that traits evolve for the good of whole species (a concept usually called "group selection"),[19] Lack saw modest overproduction and subsequent family trimming (if necessary) as a combination that could help individuals increase their personal reproductive success by allowing them to track resources. The handicapping of marginal brood members to facilitate their conditional removal was but a piece of this argument. Lack's resource-tracking concept caught on quickly, in part because it was attached to the greater issue of whether natural selection operates most potently on individuals or larger groups.[20]

All too often, a really sexy explanation tends to draw a large following, and often an uncritical one at that. Many studies of Lack's idea tended to be based on correlations (for example, baby birds in smaller families do tend to grow more rapidly than those in larger families of the same species) and the preponderance of the evidence was generally favorable at first.[21] To supporters of Lack's argument, such a trend was not surprising, but gradually cases began to accumulate that were not smoothly in accord with it.[22]

For instance, if parents were breeding as rapidly as possible it was not immediately apparent why single-egg species existed at all. It seemed as if Lack's own experimental protocol of adding a bonus egg to common swifts' nests ought to yield identical results (reduced fledging success) in every other species, too. But two early experiments along these lines produced just the opposite result. Kees Vermeer manipulated family size in the nests of glaucous-winged gulls, a species that normally lays three eggs and raises two chicks. Specifically, in Vermeer's study colony on Mandarte Island, British Columbia, unmanipulated pairs raised an average of 2.10 fledglings from their three eggs. When he added a fourth family member at 23 nests, the number of fledglings did not drop (as Lack's hypothesis would have predicted), but instead rose to a mean of 2.68. In 65 other nests, where he introduced two extra young, the average went to 3.40; and in nine more that got three extras, the average hit 4.80.[23] Where was the density-dependent trade-off?

Meanwhile, over in Scotland, Bryan Nelson was trying the same general experiment on northern gannets, a large cousin of the blue-footed booby. Like boobies, gannets make their living by plunge-diving into the ocean to grab surface-schooling fish. They lay a single-egg clutch and raise an extraordinarily high proportion of those chicks (94 percent in Nelson's sample) to fledging.[24] When he added a second chick to 13 nests, those parents fledged an average of 1.66. These results were replicated in other populations of these two species,[25] plus at least 19 others.[26] So parents of some birds are quite clearly capable of rearing *more* young than they choose to produce.

The first reconciliation of theory and data on avian clutch size took the unglamorous step of making improved measurements, essentially of having a closer look at whether parents raising extra young were, in fact, doing a satisfactory job of it. Even in its original form, Lack's point was that quality of offspring is important, so demonstrating that gannets *can* raise two chicks to the age of fledging is not quite the same as demonstrating that those chicks are as strong as they need to be. M. J. F. Jarvis repeated Nelson's experiment with a South African gannet population.[27] He got a somewhat weaker short-term result than Nelson had obtained: pairs artificially supplied with two eggs did respond with sufficient parental effort to bring both chicks to fledging some of the time (a production improvement of 28 percent over one-egg parents in one season and 19 percent in a second season versus Nelson's figure of 76 percent). But the growth patterns of "twinned" chicks left much to be desired. Though initially the siblings were the same size, one nestling in each brood tended to become considerably underweight, bringing the average fledging mass down by more than 300 grams to 2,538 grams, a difference of 12.5 percent.

Jarvis then tried to determine whether that weight shortfall was likely to affect post-fledging survival. From examining juveniles that died soon after fledging, he estimated that a critical lower limit exists, somewhere in the vicinity of 2,250 grams, and he calculated the average weight loss of fledglings that did not succeed in feeding themselves to be 51 grams per day (a conservative estimate: Jarvis actually suspected the loss to be twice that). From these figures, he projected that a typical singleton fledgling has nearly two weeks to learn the considerable skills of plunge-diving that it needs to feed itself, but a chick raised with a nestmate, being smaller when it leaves its parents' protection, has a learning period only half that long before its reserves are exhausted. He also noted that even though the heavier of two twins was nearly equal in size to singletons at fledging, the mortality rate for such chicks (over 40 percent) far exceeded that of singletons (just 3 percent).

In conclusion, Jarvis suggested that while some gannet parents *can* raise a brood of two to the point of fledging, they appear incapable of doing a very good job of it. The twinned chicks that survived the period of parental care then faced the most dangerous phase of their lives, the step into independence, under a severe handicap. He also pointed out that Nelson's data on chicks' weight, though much cited as contradicting Lack's argument that food supplies were insufficient for raising two chicks well, were few and far between by the time fledging approached. Whereas Nelson reported northern gannet twins to be essentially as large as singletons, Jarvis suspected they had probably been about 10 percent lighter.

A more fundamental refinement was made to Lack's basic theory to take into account what have come to be called the "costs of reproduction." The idea here is that natural selection actually operates on the number and quality of offspring parents produce *over a lifetime,* such that it may often pay in the long run for parents to pace themselves reproductively.[28] This should be especially important in long-lived species like eagles, pelicans, and egrets (also gulls and gannets), in which an individual that makes it through the risks of the nestling and post-fledging periods to reach breeding age (two years or more) enjoys a very long period of low annual mortality from that point on. In short, upon reaching sexual maturity such a bird has a rather large future in front of it. The potential for future breeding success that exists at any point in the lifespan (called the individual's "residual reproductive value")[29] can provide a powerful check on overexertion during a given breeding episode. In Lack's original formulation he had made the simplifying assumption that Darwinian fitness could be estimated from the current season's fledging success alone. He focused on the trade-off between quantity and quality for a given clutch of eggs, thereby missing the trade-off between the parents' current and future reproduction.

Once this point was appreciated, the time scale needed to test Lack's revised hypothesis was suitably extended. Walter Reid repeated Vermeer's classic study of glaucous-winged gulls in Puget Sound, just a few miles south of Vermeer's Mandarte Island colony, and found, once again, that parents could be hoodwinked into raising many extra young to fledging. But he also measured the wear and tear on the parents of such enlarged broods. Parents raising inflated families spent more time foraging, lost body condition, and were also less likely to survive the next winter than those bearing lighter parental loads.[30] And because a given adult's exhausted mate was less likely to survive for another round of breeding, more of the ones that did manage to make it had to start all over with a new partner, which also reduced their reproductive performance. When all the calculations were done, it appeared that the normal three-egg clutch (leading to the fledgling of two or three young) produces the highest extrapolated fitness return over a lifetime.

However, Reid's estimates also indicated that the costs associated with the next-best alternative, the four-egg clutch, were too slight to explain why that number is not standard. As noted earlier, it is always problematic to ask why something has *not* evolved, but the exercise of questioning can be useful nonetheless. If Reid's numbers are correct, then it would seem likely that the gull population would show some variation in clutch size, with three-egg types mixed in with four-egg types, a situation known as "balanced polymorphism" (familiar human examples include eye color, hair color, and blood types). But glaucous-winged gulls do not show this: they lay just three eggs.

Two possible explanations for this should probably be entertained (over and above the chance that genes favoring a larger clutch have simply never appeared). First, gulls are very adept at exploiting artificial (human) food sources, including opportunities provided by fishing boats and garbage dumps. It seems likely that modern-day gulls are living in a new and unfamiliar world in which food is much more plentiful than it was in their evolutionary past, when the three-egg clutch evolved. Three eggs may well have been the optimum number just a century or two back. Second, recall that we are just now coming to understand that the production and incubation of even one extra egg may be a lot more expensive to some birds than imagined heretofore, a point established with gulls and terns.[31] Though Reid calculated that raising a fourth chick probably lowered parental survivorship by only 2–3 percent, this was based on the tacit assumption that the production and incubation of such an addition was essentially free (when manipulating brood size, Reid moved freshly hatched *chicks,* so the receiving parents paid none of the early costs for new offspring). If those costs are nontrivial, the mystery of why there are no four-egg clutches may be solved.

Remarkably, it was not until the mid-1980s that Lack's resource-tracking idea was tested formally with simultaneous experimental manipulation of its two components, intra-family competitive asymmetries and food levels. Rob Magrath marked the population of European blackbirds in the Cambridge University Botanic Gardens. Swapping hatchlings among nests, he created 42 experimentally synchronous (size- and age-matched) broods plus 58 asynchronous (their natural condition) broods, which also contained some swapped chicks (to control for any possible recognition problems associated with having aliens among the nestmates). To explore the importance of the resource base, on some parents' territories Rob enhanced the normal earthworm diet with twice-a-day supplements of "nutritionally balanced food" throughout the nestling period. After confirming that the parents were both eating the bonus food and sharing it with their young, he was ready to test Lack's central prediction.

Supplying extra food had positive effects on growth and survival in both the synchronous and the asynchronous broods, but what mattered most critically was how the size hierarchies affected family welfare when food was abundant (supplemented broods) and when food was relatively scarce (unsupplemented broods). As Lack had predicted, when food was plentiful less hierarchical sibships did somewhat better, but when the food supply was not supplemented the asynchronous broods were considerably better off. By two weeks after fledging (the latest point at which Magrath was confident that his results were not yet affected by the juveniles dispersing), an average of 2.0 offspring from the unprovisioned asynchronous broods were still alive, compared with only 1.4 from the broods he'd synchronized. And as Lack had reasoned, the difference seemed due to the greater efficiency of brood reduction when siblings were unevenly matched. Midway through the nestling period, 64 percent of the asynchronous broods experiencing "poor food" had already lost chicks, while in comparable synchronous nests 86 percent of the broods were still intact, with parents still pouring investment into chicks that would soon die. In the end, survivors of brood reduction ended up 12 percent heavier in the asynchronous treatment.[32]

A more recent experimental test of resource-tracking was conducted by Scott Forbes's group in Winnipeg, following up on their insurance test with redwings, again making the point that these two incentives for parental overproduction are mutually compatible. Before manipulating broods, they established that density-dependent growth is the rule and that the day-later marginal chicks pay the price when food is short. Specifically, on the fifth day after the core brood hatched, nests with heavier core chicks also had relatively heavy marginal chicks, but nests with slightly underweight core chicks had *very* underweight marginal chicks. That is, when the family budget is tight, the marginal chicks' rations are cut first. This suggests that the system is reversible so long as the marginal chicks remain alive.[33]

To examine how family size and tight budgets affect core and marginal brood members, Forbes's group conducted a six-summer experiment near Winnipeg. Because growth and survivorship varied widely from year to year, they analyzed food availability indirectly (the three seasons with above-average survivorship categorized as "good" years; the three below-average seasons as "bad" years) to test a key prediction from Lack's hypothesis. If marginal eggs produce bonus offspring when food is relatively plentiful, then the fates of marginal chicks should be far more sensitive to both family size (intensity of demand) and poor food conditions (magnitude of supply) than is the case for their core-brood nestmates.

This sensitivity was expressed in terms of the disparity between mortality rates of core and marginal chicks, which varied with both degree of crowding and richness of season. In experimentally enlarged broods, for example, the mortality rate of marginals was 43 percent higher than that of core nestmates in low-food seasons, but the difference was only 23 percent in good years. In experimentally reduced broods the same basic pattern was evident, but on a much lower scale (differences of 22 percent in bad years and 2 percent in good years). Sham-treated and unmanipulated controls were intermediate. And in normal-sized broods that happened to experience some hatching failure, essentially becoming reduced on their own, a similar but modest shift occurred (32 percent in poor years; 13 percent in good).

Thus the two likely avenues for gaining relief from competitive squeezes, increasing the supply and decreasing the demand, both had their expected effects. When food is scarce, marginal chicks bear most of the burden; when it is more plentiful, they can become full citizens. These results put a slightly different twist on Lack's original perspective. Because some marginal chicks die under even the most favorable conditions, and because marginals achieve the lofty status of core chicks only when a core sibling is lost (via hatching failure or experimental removal), the parents' apparent "goal" for their progeny is likely to be somewhere below full-brood success. Not knowing what conditions will arise in the upcoming breeding season, parents open with the modest ante of a single egg, an amount they can probably afford to lose. We turn next to consideration of what offspring can do, if anything, about their parents' gambling addiction.

10 Beggars, Cheats, and Bad Fruit

When the only tool you own is a hammer, every problem begins to look like a nail.

—Abraham Maslow

Nestling birds often react frantically when parents arrive with food. They vocalize shrilly, move clumsily about the nest, and do other things (such as egrets' scissoring of the parent's bill) that are lumped under the collective label of begging. These actions have long been suspected of serving both to inspire the parents toward continued (and perhaps accelerated) food service and also to make each beggar's case for why it should be the one to receive the next mouthful. Both of these proposed functions can be framed as sibling competition. If begging induces parents to work harder than they otherwise would so as to deliver more food, the extra effort may reduce their ability to create and invest in future broods (siblings of the beggars): parents can suffer burnout. More immediately, if a given beggar's performance increases its likelihood of getting this particular worm, then that meal is obviously not going into the gullet of a nestmate.

Signaling between dependent young and their caregivers has attracted increasing attention from researchers because it combines two theoretical themes: parent-offspring conflict, which we have encountered throughout this book; and communicating in general. To get a feel for this second theme, consider a simple dyad in which the players are the sender of a given signal and the receiver thereof. The sender has some information it wishes to convey, which it encodes into one or more forms of energy for the trip over to the receiver's sensory system. As I write this paragraph, for example, I am encoding my thoughts via tactile energy through a keyboard and into my word processor, but the receiver is pretty remote to say the least. Eventually, I hope that some receivers will take in my thoughts as patterns of reflected light from a page. A moment ago, I used vocal energy to communicate to our dogs. While jogging this morning, I used reflected light energy to wave to a neighbor, and so on.

Early animal behaviorists assumed that the act of signaling was all about the efficient and accurate transfer of information. The process was regarded as intrinsically cooperative, in the sense of being mutually beneficial to both parties. And certain mating signals were especially impressive for their efficiency. For example, in many insects the female emits a minuscule volume of odorant (highly specialized organic compounds called pheromones) and downwind males as far away as a mile or more may get the message. Male moths, for example, have large, feathery antennae. Each species' antennae are extraordinarily sensitive to the species-specific odor produced by its females. The males obey a simple behavioral rule of "Fly upwind till you lose the scent, then circle around."

A quarter of a century ago, the topic got a conceptual overhaul from the proposition that animal signals exist only because one of the two players, the sender, wants it to exist.[1] (I am using the term "signal" to include only voluntary acts of communication, like intentionally squirting perfume from a gland.) When I wave to my neighbor, it is something I can either do or not do. The reflected light that he perceives is affected by my arm motion, and his nervous system can then do with that information what it will. For the great majority of animal signals, in fact, a benign exchange of accurate information may be a correct description. Both the male and the female moth need to find each other for sex, so the shared benefits seem clear. In other cases, the consequences may be negligible: by waving I signal that I see and recognize my neighbor, and perhaps that I am a friendly sort.

But the cynical insight about why signaling exists can accommodate less genial signals as well. We needed a framework for understanding that some signals may be given because the sender wishes to manipulate some aspect of the receiver's activities in ways that may not be good for the receiver. For example, it has been proposed that nestling birds' begging may interact with predation pressure to blackmail parents into providing more food.[2] Because begging is noisy and many predators have ears, it is possible that offspring coerce their parents into keeping them happy as a means of keeping them quiet. There is even a bit of experimental evidence showing that begging noise can jeopardize the contents of ground-nests.[3]

Indeed, some signals may be downright disastrous for the receiver. A sinister example will make my point. Fireflies (a.k.a. lightning bugs), those familiar icons of summer twilight, are conspicuous signalers. We notice them because they generate their own light chemically and can turn it on or off at will. Most of this flashing (pun intended) is performed by cruising males as they fly around trying to make contact with receptive females of their species. Object reproduction. Extraordinary fieldwork by Jim Lloyd of the University of Florida revealed much about how their signal system works.[4] After inventing a special fishing rod with a tiny light bulb hanging from its tip and the on-off switch back at his thumb, Jim could "be" a simulated firefly himself and thus investigate which aspects of the signal were important to the various players. He found that each species of the common genus *Photinus* has its own particular code of male flash and female counter-flash. In species X a male might turn his light on for a fixed time, say 1.2 seconds, while flying in a particular pattern, making the letter "J" appear as if by magic in the deepening gloom. A female of X perched on a grass-blade nearby sees his pattern, waits a precise interval, say 0.6 seconds, and gives a response flash of the appropriate female duration. The male descends to the spot of her flash and they get it on. In nearby species Y, the timing and flight pattern are a bit different, leaving little chance that two genetically incompatible mates (different species) will copulate with each other. So far, this seems like a signaling system that is mutually beneficial for both parties, regardless of species.

But there is a third party, a predatory genus *(Photuris)* whose females have broken the lovers' code and use their own flashing ability to play the siren, luring male *Photinus* down to a fatal roll in the hay. Amazingly, individual femme fatale fireflies can impersonate several *Photinus* species on the same night. The reason males go to their death should be obvious. These guys do not live very long in any case, so they are in a hurry to reproduce. Besides, the great majority of tiny flashing lights that reply with the perfect password are, in fact, desirable partners. In addition, he who hesitates may lose out to other males eager to claim the sexual favors of *Photinus* females. So the flames of male haste are fanned by the twin frequency-dependent realities of the rarity of femme fatales and the plenitude of male rivals.

This is clearly another evolutionary game. Natural selection acts on signal senders to refine their sales pitch while simultaneously acting on receivers to be a bit skeptical and sometimes downright resistant. Male *Photinus* fireflies do rush down to the ground toward a female's flash, but they often make their final approach with considerable caution.

From this perspective, then, we can consider what information is really encoded in the vocalizations and eager gaping acts of begging nestling birds and whom it benefits. Robert Trivers raised this issue explicitly in his classic paper on parent-offspring conflict, wherein he proposed that natural selection acts on youngsters to try to skew parental investment toward themselves.[5] He reasoned that they might well use deceit, overstating their true physiological needs, and that the parents might not be sufficiently well informed to make the best possible corrections. Accordingly, the matter of whether offspring are dishonest lies at the heart of much current study. And the fundamental idea of offspring signaling as a dynamic social game, with various players adjusting their tactics in response to changes in food deliveries and what their rivals are doing, has led to some fascinating developments.[6]

For example, field experiments with American robins have shown that nestmates are highly attuned to one another's begging levels. The idea is that chicks solicit food with varying levels of vigor, thus can adjust their behavior in accordance with how valuable the next food item is and how high a personal cost they are willing to pay for an enhanced likelihood of getting it. Usually the chick that begs the fastest and loudest and/or has positioned itself most favorably is the one that gets fed.[7] So if one chick is removed from its nest and deprived of food for a while (kept warm and well cared for otherwise), it pulls out all the stops when returned home and, on average, gets more of the next few deliveries. And when one chick is thus manipulated into raising its ante, its undeprived nestmates escalate their begging also.[8] The game is intensified.

This result has been replicated with European starlings, but using a subtler protocol. Alex Kacelnik and his colleagues at Oxford placed a chick either in a family larger than its own (thus stiffening the competition it faced), or in a smaller brood (for the opposite effect), or in a same-size brood (as a control). After a period of living in the substituted competitive milieu, the chick was returned to its home nest and observed. The results show clearly that chicks recently exposed to keener rivalry for food beg harder and get more food, while those that have had it relatively easy give less effort and reap fewer rewards in the short term.[9] Subsequent work with barn owl broods suggests that a sigificant portion of chick begging may actually be aimed *at* fellow nestmates and not just parents, with rivals communicating their own willingness to escalate competitive efforts over the next food item.[10] That is, just as fighting among siblings lets a subordinate know what competing will cost it in terms of punishment, nonviolent signaling may provide information about how much various rivals are prepared to escalate in other ways to secure undelivered food. This idea also helps account for an otherwise puzzling proportion of nestling begging signals that are given when no parent is present, that is, when only fellow nestmates are available as signal receivers.

Another experimental ploy for manipulating begging signals was developed by Barb Glassey, who applied topical anesthesia to the vocal apparatus (syrinx) of nestling red-winged blackbirds, rendering them temporarily mute. In a nest where all four chicks were gaping and stretching but one of them was producing little or no sound, parents gave less food to the silent one. (This short-changing of the weak-signaling nestling did not translate into a higher allowance for its noisier siblings, though, because parents simply kept more of the food for themselves.)[11]

Although escalated begging generally improves an individual chick's take, the playing field is not necessarily level. These dynamics of competitive soliciting have been studied with nestmates unequal in size and age. In several species it has been shown that junior sibs beg harder than their senior nestmates, but get less. When the hunger levels of yellow-headed blackbird chicks were manipulated experimentally, their begging varied as expected (stronger when hungry), but the resulting food allocations remained decidedly uneven. Like young robins and starlings, yellow-headed blackbird chicks escalate begging when in the presence of a hungry, begging nestmate.[12] So a parent returning to a nest with one very hungry (food-deprived) junior chick confronts several active beggars, most of whom are larger than the deprived one and likely to be more effective at competing. At such times, parents are inspired to make more deliveries, but the extra-hungry junior sibling still gets less than its share.[13]

Parent birds apparently can discriminate between their larger and smaller offspring, as revealed by a few species in which preferential feeding of the smallest chick has been reported. That is, offspring size and motor skill disadvantages can be overruled if parents take the time and trouble to sidestep the big guys and see that the food gets to the runt. In captive budgerigars, for example, fathers feed whoever meets them at the nest-box opening, basically just shoveling food through the hole without apparent concern over who gets it, but mothers carefully locate the smallest chick for feeding.[14] Similar behavioral favoritism has been reported for a handful of other birds,[15] and the reverse, preferential feeding for larger offspring, has also been reported.[16] To the degree that parents have the ability to assert their preferences, then, we cannot be sure whether the most commonly observed distribution of food, namely some degree of skew toward larger siblings, is due to parental decisions, to effective sibling competition, or to both working in concert.

A parent reed warbler flies to the rim of its cup-shaped nest, a freshly caught insect in its bill, and confronts the begging brood. Its four nestlings respond to the arrival by raising their small heads up to the highest extent of pipestem necks and uttering a cacophony of "si . . . si . . . si . . . si" calls. The parent stuffs the food into one throat and departs again at once, presumably carrying some impression of the chicks' current states in its memory.

Becky Kilner and her colleagues sought to decode the complex display of begging in reed warblers by analyzing which aspects of the multimedia stimulus package actually influence parental deliveries. By videotaping and painstakingly analyzing many meals, they zeroed in on two specific cues, one visual (the area of the total brood's combined gapes) and one acoustic (the rate of all four chicks' combined vocalizations). Other factors, such as the bright yellow hue of the mouth linings, did not seem related to the number of parental deliveries per hour, but these two main signal features accurately predicted most of the ups and downs in parental performance. In fact, the acoustic cue seemed to shape somewhat different aspects of the parents' behavior than did the visual cue, and the combinations of these components apparently gave more refined information about offspring states. Still, making sure that the observed fit between the chicks' signal strength and the parents' responses was due to a causal relationship—and not merely to variation in some unnoticed other factor(s)—required a properly controlled experimental approach. A series of field experiments followed, including such steps as seeing how food deprivation affects the chicks' performance of the two signal components, and how one or four extra voices of beggars (added by tape playbacks through a small speaker attached to the nest) influence parental deliveries.

Because young reed warblers are engaged in a fantastic growth spurt, their displays strengthen a bit each day. Parents accommodate the modulations in signal form by integrating the complex of acoustic and visual stimuli through some unknown physiological mechanism that can be, and has been, rendered mathematically as a regression equation.[17] The details of that contrivance need not concern us here; what matters is that there is a flexible "simple behavioral rule" that enables parent warblers to understand their developing young.

This complex and shifting signal appears to be "honest," that is, to report accurately what the brood's needs are. Kilner's team performed experiments with chicks that had been removed to the lab (and deprived of food), then stimulated to beg at set intervals by light jiggling of the nest; the chicks' responses grew stronger as their hunger (their time without food) increased. Similarly, when chicks were deprived of food for varying periods and then fed to satiation, the amount they ingested varied both with their presumed hunger and with the level of their begging signals.

Thus, it appears that, at least in this species, chicks may be giving their parents truthful information about their needs. They do not seem to be deceiving the parents into doing more work than is in the parents' best interests. Nevertheless, I remain quite skeptical about just how honest begging birds really are. My doubts center on two points. First, few if any of the empirical studies have addressed the essential matter of offspring *need* in a clear and convincing manner, opting instead for the much more convenient surrogate we call *hunger.* As we all know, desires and needs are not necessarily synonymous. (I may *want* a Ferrari, but it would be hard to make the case that I really need one.) Second, there is substantial evidence that some parents certainly do not direct their discretionary investment toward the "neediest" offspring. On the contrary, there is abundant reason to believe that many parents show favoritism toward their strongest and most robust offspring, which is directly counter to the current versions of the honest-signal argument.[18] We just saw this in the case of yellow-headed blackbirds, where parents commonly give food to the largest nestlings, even when the smaller siblings are begging more acutely. Magellanic penguins (discussed in Chapter 6) provided another clear example. I shall return to the topic of parental favoritism shortly.

On the surface, though, reed warbler parents and offspring seem to be generally cooperative with one another. Parents receive accurate information about hunger, at least, from the chicks' begging signals and are responsive when more food is requested. While one can always wonder if there is really *no* disagreement over the details, the basic picture seems to be one of harmony.

However, just as in the firefly system, there is a sinister additional player in this drama, one that changes the whole picture. The common cuckoo is a brood parasite, which means that it does not build its own nest or raise its own young. Instead, the mother cuckoo lays her eggs in the nests of various other species, one per nest, and lets the host parents do all the work of raising young cuckoos. For this purpose, the parasite has evolved many famous specializations. First, each cuckoo egg is held inside the mother's body for an extra day, where it receives constant warmth that allows development to begin immediately. This day is subtracted from the time it will take the host parents' incubation to finish the job, so hatching occurs that much more quickly after laying. Second, cuckoo eggs have relatively thick shells, which are believed to help them survive the short fall from the mother's body to the floor of the hosts'

An adult meadow pipit feeding a fledgling common cuckoo. The host species provides incubation and parental care to the brood parasite, which kills the host's own young by evicting them forcibly from the nest. Because the cuckoo is not related to its nurserymates, there are no kinship bonds to temper its selfish behavior. (Photo: M. Strange.)

cup nests (which the female cuckoo is much too large to enter). This falling missile may crack any host eggs already in the nest. Third, by emerging from its egg very early, the cuckoo chick has an opportunity to clean house. Naked and with eyes still closed, it seeks the smooth hard surfaces of any unhatched eggs or the soft, warm bodies of freshly emerged hatchlings. Turning around, the young cuckoo backs into its target, pushing it up the nest wall, balancing the victim between tiny wing stubs, until the victim falls over the rim to its death. The interloper continues this lethal activity until all the host's eggs or chicks have been evicted.

But why do hosts go along with this outrage? This is one of the great puzzles in natural history. Why does the warbler parent not simply lift or push the cuckoo egg out of the nest before it hatches? Failing that, why does it not take one look at the ridiculously large chick, recognize it as nobody's idea of a reed warbler, and cut its losses by deserting the nest? Does natural selection not act on the DNA in warbler chromosomes?

One could write a whole book on the cuckoo-host arms race, but I shall restrict myself to three quick partial answers.[19] First, recalling the dilemma faced by *Photinus* firefly males, it is clear that not all reed warbler parents have to face the cuckoo challenge. In Wicken Fen, Cambridgeshire, where Kilner's study was done, the annual rate of parasitism varies from 1–2 percent in some years up to nearly 25 percent in others. This means that there is a probability of between .75 and .99 that a given nest does *not* have a cuckoo problem, which provides substantial rewards to parents using the simple convention of accepting whatever they find in their nest. Reed warbler nests almost always contain reed warblers. To make this frequency-dependence argument another way, 100 percent of all cuckoos alive today descended from successful exploitation of a host species, but the hosts probably never faced a selection pressure nearly that strong. Second, it is apparent that hosts do sometimes identify and remove cuckoo eggs, but this inexorably leads to ever-fancier counter-strategies by cuckoo genes. The common cuckoo long ago diverged into a handful of sub-populations, called "gentes," each specializing in a single host species. Each of these gentes has had its eggs disguised by the slow hand of natural selection so as to match the color, size, and patterning (speckles, swirls, whatever) of the host's own eggs.

It has also been argued that reliance on learning to recognize proper chicks by appearance could easily backfire on host species. That is, if a reed warbler parent imprinted on its first brood to make a lifelong template of what a reed warbler nestling should look like, it could be ruined if it chanced to have a cuckoo in its first brood, because then any real offspring it ever managed to hatch would look wrong and be rejected.[20]

Whatever the explanations, the fact remains that host parents hustle back and forth, stuffing food items into the parasite as it begs. So the matter of parent-young communication returns. The visual component of the offspring signals to which warbler parents are so attuned is clearly out of whack—not just the gestalt of an immense nestling, but the salient detail of gape area. Despite being far larger than any warbler chick, the cuckoo's bill is certainly not equal to the summed area of the gaping mouths of four warbler chicks. Kilner's research team wondered if there was something else special about the cuckoo's total signal package that inspired parental effort, so they presented host parents with a single cuckoo-sized nestling belonging to a completely different species, namely a thrush or a blackbird. Warbler parents were willing to feed such unfamiliar guests, too, but not with the zeal they showed for cuckoos. The vocal signal produced by a young cuckoo offered a possible key to the problem: the cuckoo emits a rapid "si, si, si, si . . ." that actually sounds like a *whole brood* of warbler chicks. Indeed, at age one week it matches the vocal output four warbler nestlings, and a few days later it sounds like eight! When the otherwise unremarkable thrush chick was accompanied on tape by the sounds of either a whole brood of warblers or a single cuckoo, it too was fed at full speed.[21]

Apparently what has happened, then, is that natural selection has shaped the cuckoo's calling to compensate for its inevitable lack of visual clout. There must have been many false starts and dead ends in the evolutionary trial-and-error process, but the end result is a jury-rigged combination of elements that manages to fit well enough into the warbler's preexisting sensitivities to satisfy the integration rule that parents use in governing their effort level. So the host adults deliver as much food as they would normally bring to their four young, and, because cuckoos fledge at a later age than warblers, they even extend this room service ten days longer than normal.

Brood parasitism creates the quintessential dysfunctional family. When a baby cuckoo hatches, its potential for acute competition over limited resources is not checked by the social glue of relatedness. To be sure, the common cuckoo is an extreme, a caricature of how far down the road to ruthless exploitation such a system can move. There are less drastic but instructive compromises, variations on the brood parasitism theme. In fact, part of the young cuckoo's problem results from its own actions: if it did not throw out all the host offspring, they could help with the visual signaling. One presumes that host chick "assistance" was less valuable to cuckoo chicks than their absence, but that is not always the case.

A less obnoxious parasite, albeit still no treat for hosts, is the brown-headed cowbird, a smaller relative of the red-winged blackbird, which parasitizes more than 240 species of North American songbirds. When the female cowbird is ready to lay, she slips into a host nest, removes one resident egg in her bill (puncturing its shell if it is large, gripping it intact if small), and quickly leaves her own egg in its place. In host species bigger than the cowbird, the chief cost of being parasitized can be this removal of an egg, since cowbird hatchlings do not evict or otherwise assault their nestmates and bigger host chicks can compete reasonably well with the intruder.[22] A South American species, the shiny cowbird, operates similarly with large host species except that the laying female punctures one egg without removing it.[23]

Smaller hosts of cowbirds, by contrast, often lose some of their young to post-hatching competition; and if their incubation period is longer than the brief ten to twelve days needed by cowbird chicks to hatch, they commonly lose everything. In such nests, the cowbird chick usually receives more than its share of the food.[24] This raises the same questions: Why do host parents not only tolerate the intruder's presence, but actually allow it to eat the most? Are the parasites simply better competitors? Or is there something more?

In the early days of classical ethology (1930–1960), Konrad Lorenz and others noticed that many animals show behavioral preferences that are bounded on the lower end (that is, requiring some minimal amount of stimulation to produce a response) but apparently lack an upper limit (ostensibly because "too much" of that particular stimulus seldom occurs in nature, so there are no penalties associated with having an open-ended sensitivity). This was dubbed the "supernormal stimulus" concept, and it produced many memorable experiments. For example, some birds prefer large eggs to small ones. It turns out they like absurdly large eggs even more. If given the opportunity, such parents will roll a fake egg the size of a football into their nests. The thought is that somewhat larger eggs are often more valuable (having greater nutrient stores) than small eggs, hence the original sensory bias.

If super-sized egg lookalikes were common in the real world, natural selection would be expected to favor evolution of an upper limit. So the puzzle with host parents lavishing attention on gigantic, extra-loud, and cheeky brood parasites comes down to this: Is such a nestling a true supernormal stimulus that the host parents are unable to resist? This tantalizing possibility, that host parents are duped into favoring the intruder, rests on the cowbird chicks having one or more special features that make them more appealing to the host parents than the parents' own genetic young. In the particular case of cowbirds, the puzzle is complicated by the fact that they are extreme generalists as brood parasites: how could a refined bit of "psychological weaponry" evolve that works on hundreds of different host species? Might a single key open all those locks?

Gabriela Lichtenstein from Cambridge and Spencer Sealy from the University of Manitoba focused on the interactions between brown-headed cowbird parasites and yellow warblers, small, fast-incubating hosts that suffer only partial brood losses from parasitism. At Sealy's long-term site on the great Delta Marsh in southern Manitoba, an average of 21 percent of warbler nests are parasitized by cowbirds.[25] Fifteen broods, standardized to contain just one warbler and one cowbird chick, were videotaped for five two-hour samples during the period of peak food demand (ages 3–5 days). The tapes were scrutinized to ascertain whether host parents actually showed any behavioral preference for the parasite over their own offspring, the hallmark of the supernormal hypothesis. As it happens, they did not: warbler adults delivered without any detectable prejudice to whichever chick happened to be tallest at the moment of the parents' arrival. This is good news for yellow warblers, but still not great, since the slightly larger cowbird chick tended to achieve that position more often. In short, cowbirds appear to out-hustle, out-beg, and out-stretch their warbler nestmates.[26] Still, this deal has to be better than being shoved out of the nest.

Not all brood parasites are birds. An African cichlid fish that performs a form of parental care called "mouth-brooding" is subject to invasion of the courtship nest by a local catfish. Cichlid fish eggs are fertilized externally and then taken into the mother's mouth, where they are protected both as eggs and later as fry. The catfish sneaks its fertilized eggs in among those of the host during the brief spawning moment before the egg mass is transferred to the host's mouth. Like young cuckoos, catfish fry are hostile guests, hatching sooner and eating some of the eggs and hatchlings of the host species.[27] Ants are also prolific practitioners of the parasitic arts.[28]

To close this treatment of brood parasitism, let us consider an unusually low-impact version of the phenomenon. Because ducklings are highly precocial, hatching with full coats of down and advanced abilities to walk, swim, and feed themselves, a mother working alone can provide all necessary care relatively easily. She leads her brood to areas of safety and rich food, remains alert for sources of danger, and offers a modest level of defense (charging small predators and using broken-wing deception to lure larger ones away from the well-camouflaged chicks). The Barrow's goldeneye is a medium-sized sea duck common along the northwestern coasts of Canada and the United States. The reddish-headed female lays seven to ten eggs in a tree cavity (or nest box) and incubates them alone for a month. As with other ducks, hatching is highly synchronous, and the whole brood jumps down from the nest cavity one to two days after hatching to spend the ensuing two months paddling around freshwater habitats with the mother before dispersing off to sea. Of interest to us is the fact that ducklings belonging to one mother often show up in the entourage of a different female. They are sometimes accepted and treated like family; other times, driven away. These two opposite reactions to the adoption process raise questions about whether these intruding ducklings have a harmful impact on resident brood members. If they do, why allow them to tag along; if not, why bother driving them away?

In a pond-studded area of British Columbia, John Eadie and Bruce Lyon trapped mothers on their nests, color-banded them for identification, and kept close tabs on their clutches so as to be present at the moment of hatching. Then they marked the ducklings with small plastic tags in the toe webs and waterproof ink marks on their cheeks. Knowing who belonged initially to whom is what made post-hatching rearrangements detectable. The hypothesis under study was that brood amalgamation stemmed from three factors: (1) the tendency of mothers to desert undersized broods (on the logic that the risk to the mother's reproductive future associated with staying with ducklings may not be sufficiently compensated by the opportunity to raise only a few offspring); (2) the desperation of abandoned ducklings to join any brood; and (3) the indifference of (amply compensated) mothers to having extra ducklings in tow. To test this argument, broods were experimentally reduced (to explore point 1) or expanded (point 3).[29]

Twelve goldeneye broods were experimentally reduced just after hatching to just three or four ducklings, then compared with the fates of fourteen other broods that were handled but not reduced (they averaged nearly ten ducklings). Nearly all the control females remained with their broods, with just one abandoning and another whose whole brood simply vanished. In the reduced broods, by contrast, just four (one-third) of the mothers remained with their broods, while the other eight abandoned their young or disappeared. This difference in maternal response is (statistically) unlikely to have occurred by chance, so it appears that point 1 of the hypothesis is borne out: mothers of reduced broods really are more likely to abandon their offspring.

Deserted ducklings are in pretty dire straits. If they remain on their own, the chance of surviving is only about one in twenty, and if they can join another brood it jumps up to one in two. So point 2 also seems in order. But the adopting mothers presented an interesting wrinkle. When Eadie and Lyon added ducklings to broods, mothers were graciously accommodating if the intruder was very young and were quite the opposite (even murderous in two cases) for chicks older than about ten days. This reversal of attitude may be due to a mother simply needing some time to learn all her own offspring as individuals: when broods are very young their mothers may not be able to discriminate which do not belong. Or perhaps younger additions to a mother's brood are an asset (offering alternative targets to predators during the period of highest risk), while older ones compete too strongly with her own ducklings for food and other resources.

Another form of signaling between offspring and other family members is so cryptic as to be easily overlooked, but it is subject to the same evolutionary pressures and principles. While still attached physically to its mother, a mammalian embryo exchanges materials in both directions, nutrients in and wastes out, across a highly specialized tissue interface, the placenta. Something quite similar occurs between an apple and the limb to which it is attached. In pipefish and seahorses, it is the father (and his brood pouch) that plays the parental role, with material exchange occurring by diffusion. In all three of these systems, the level of metabolic activity within the dependent offspring is likely to affect the rate at which additional nutrients are supplied, either actively (by passing particular hormones out as signals to the parent) or passively (to the degree that consumption of resources creates a vacuum that is then filled by the provision of replacement nutrients).

We can examine these physiological relationships in much the same way we analyze begging behavior. For example, there are likely to be many cases in which it is in the parent's best interest to increase investment in an embryo that is growing very well (insofar as doing so helps produce grandkids) or to truncate investment in one that is faring poorly (to avoid throwing good investment after bad). This would be a simple physiological analogue of parent birds that shower extra care on their larger offspring and ignore (or eject) the weakling. Accordingly, in mammals and plants we should expect to find various forms of spontaneous abortion that are based on assessment of which embryos are performing well, even as the parents also resorb or abort portions of the initial crop of offspring in response to current budgetary constraints. A mammalian mother is more likely to miscarry all or some of her litter if she is underfed; a plant that gets too little rain tends to drop its flowers and set fewer fruit. A plant is also more likely to jettison offspring that are not well served by income-generating leaves (perhaps you pruned those particular leaves) or that are on a side of the plant that gets less water. As Shakespeare wrote in *The Merchant of Venice,* "The weakest kind of fruit drops earliest to the ground."

There may even be a potential for certain kinds of misinformation being telegraphed from embryo to parent as a means of skewing extra investment to the embryo's Self. Here I allude to possible evolutionary conflicts of interest. To quote W. D. Hamilton, a modern theoretical bard, "for one seed to expand selfishly at the expense of its neighbours may or many not be advantageous to the inclusive fitness of its genotype, but it is almost certainly not in the interest of the parent plant."[30]

We are now learning of a phenomenon called "genomic imprinting" that seems to be a manifestation of such conflict. In plants and mammals whose females receive pollen or sperm from multiple males, offspring are likely to be surrounded in the nursery by maternal half-siblings, which is to say with rivals bearing some shared alleles (derived from the mother) and some unshared ones (derived from their different fathers). But the mothers have to pay all post-fertilization expenses. Genes that arrived in sperm or pollen thus have an opportunity to further their own replication rates by inducing the body they're building to consume limited resources at a very high rate, since the cost of their selfishness is most likely to be borne by neighbors not carrying copies (because fertilized by different males). From the perspective of such alleles, the next embryo over in the womb (or on the limb) is no more a relative than a reed warbler nestling is to its cuckoo nestmate. Without imputing any cognition (or even recognition) to genes or embryos, we can see that there is an avenue of reward open to a selfish trait that is paternally conferred. Such genes, expressing themselves as voracious consumers only when transmitted in sperm, have been found in mice[31] and in corn.[32]

The physiological signals that genomically imprinted genes appear to be directing toward mothers are not imagined to be "honest" indications of true need. Like the begging of a cuckoo chick in a reed warbler's nest, they are exaggerations, manifestations of selfishness untempered by inclusive fitness incentives for altruistic treatment of close kin. (The word "cuckold" reflects this parallel between avian brood parasitism and the introduction of alien offspring into someone else's family.) We shall turn next to broader considerations of whether offspring generally can manipulate their caregiving parents.

11

Silly Squabbles and Serious Sabotage

Remember the generational battles twenty years ago? Remember all the screaming at the dinner table about haircuts, getting jobs and the American dream? Well, our parents won. They're out living the American dream on some damned golf course in Vero Beach, and we're stuck with the jobs and haircuts.

—P. J. O'Rourke

We have seen how selfishness is expressed between siblings, starting with unalloyed assault and moving down to competing signals from offspring to the parents who feed them. Now the focus shifts more fully to the interaction between generations. It is clear that the traditional view of parents simply pouring investment into passive offspring was inadequate for a variety of reasons, and Robert Trivers's highly influential theory of parent-offspring conflict (introduced briefly in Chapter 4) merits further exploration.[1] To summarize Trivers's argument briefly, the fitness of parents is assumed to be maximized *when their investment is allocated equally* among all their offspring, a deduction resulting from the fact that the coefficient of relatedness between parents and offspring (in sexual diploids) is always the same, $r = .50$. Parents are therefore presumed to value all offspring equally. By contrast, each offspring should view its own personal fitness as being twice as valuable (making twice the contribution to its inclusive fitness) as the personal fitness of even a full sibling, here because of asymmetrical relatedness (r to Self $= 1.0$, r to sib $= .50$). Thus a given unit of parental investment that goes to Self instead of to a sib should be regarded by the selfish offspring as having an attractively halved cost. From this relatedness logic, then, it follows that natural selection should shape offspring traits toward higher degrees of selfishness than it shapes parents' willingness to tolerate.

Noting that nursery-age offspring are virtually always at a great physical disadvantage in this intergenerational war of wills, Trivers proposed that they must use their ace in the hole, namely superior information about what their true needs are. In particular, they should inflate their signals of need so as to take advantage of the parents' inability to distinguish needs from wants. The desperate-sounding wails of human infants could thus be seen as ridiculous exaggeration, but belonging to a class of bluff that parents cannot call without risk. Just as fire companies must pile onto the hook-and-ladder truck and race to every signal of disaster, knowing full well that some of these are false alarms, so might parents have to allow themselves to be manipulated: they cannot take the chance. Obviously, this ties into the issue of whether youngsters are honest in their communications to their parents.

Trivers proposed these two points—asymmetrical relatedness plus the weapon of deceit—as the basis for explaining many rancorous aspects of parent-offspring interactions. He focused much of his attention on "weaning conflict" between mammalian mothers and offspring, arguing that the lactating mother will favor an earlier weaning (the better to save herself for production of future babies) than the current infant (which is getting a rich diet without having to do much). Anyone who has ever been around a litter of puppies or kittens can recall the persistent clamoring of offspring wanting to nurse and the escalating withdrawal behavior of the mother. For that matter, because all of Trivers's readers had been children at one time and many of them were enduring the trials of parenting, his whole explanation struck a very deep chord. This piece of theory seemed to shed considerable light on everything from tantrums in grocery store checkout lines (where store managers fan the flames by parking gleaming rows of candy right where basket-riding toddlers will be forced to sit and contemplate them) to the shrill importuning of nestling birds.

The theory of parent-offspring conflict also had the irresistible appeal of being simultaneously unexpected and deliciously simple. It deserved that highest of accolades in theoretical biology, the status of being "elegant."[2] And, as described earlier, it faced an immediate firestorm of criticism before receiving the seal of approval from mathematical testing of its basic logic. As the years rolled by, the theory of parent-offspring conflict became a fixture in the firmament of behavioral ecology, and when Richard Dawkins brought out a new edition of his 1976 classic, *The Selfish Gene,* he added an endnote that concluded: "There is indeed rather little to add to [Trivers's] paper of 1974, apart from some new factual examples. The theory has stood the test of time."[3]

This is precisely the point I examine in this chapter. Considering myself to be a serious fan of the theory, and convinced that the family-as-battleground argument itself is sound, I have nonetheless come slowly to question whether there are many, or possibly any, "factual examples" of traits that have evolved from the particular forces identified by Trivers. That is not to say that the theory is merely a heuristic bauble with which biologists can amuse themselves. It has led to many theoretical extensions of the logic[4] and great progress in related directions (such as the burgeoning literature on whether nestling begging is honest),[5] but a keystone theory should, it seems to me, provide a unique explanation for phenomena that either cannot be understood any other way or at least cannot be understood so well. My concern, then, is not with Trivers's gem of an idea (I wish I had had it!), but with myself and my fellow empiricists, who have made too little real progress with it in the decades since 1974.

Having breathed that heresy let me hasten to concede a few crystal-clear cases where the evolutionary interests of parents and their offspring are surely incongruent and where one side prevails at the expense of the other. As shown earlier, egg size correlates positively with subsequent offspring success in various birds. It follows that any mother laying an egg of below-average volume or content, or for that matter even below the largest she could possibly afford to produce, is performing an act that is presumably in her best interest (allocating her resources optimally across two eggs, for example), but at some cost, and quite possibly at a net inclusive fitness cost, to the embryo inside the smallish egg. *Et voilà,* parent-offspring conflict.

We have already seen other examples of intergenerational contentiousness: in two previous chapters, we encountered studies in which parent birds (cattle egrets and red-winged blackbirds) cut their food deliveries to hungry broods that were manipulated experimentally (via chick-removal and single-chick muting, respectively). A similar effect has been reported for brown pelicans.[6] It was likely (from their continued begging) that the chicks were still hungry (and in the blackbird case, all mouths were still present) and unlikely that parents temporarily forgot how to hunt, so it looks very much as if parents sometimes just keep more of the family's groceries for themselves.

Yet another example comes from long-term research on red deer on the Scottish island of Rhum, where Tim Clutton-Brock and his colleagues from Cambridge have been following individual case histories since the early 1970s. A fawn begins life as its mother's only current offspring and nurses for 30 weeks, during which time its weight climbs from 15 pounds at birth to 90–110 pounds. At this point the mother rejects its attempts to suckle, forcing it to graze on local grasses, a change that allows her to resume estrous cycling in time for the autumn rutting season. The mother's decision to wean the calf then is in her best fitness interests (that is, the next pregnancy makes a sufficient contribution to her lifetime reproductive success to compensate for the benefit her current calf could derive from continuing to suckle). But the field data also show that early weaning is not in the calf's best interests. If the mother fails to conceive, the calf is allowed to resume nursing, and such calves get through the winter more often and in better condition. So here we see a nice conflict in fitness interests for the two players: the calf's fitness might well profit more from getting the milk than from gaining a sibling (or more likely a half-sibling, as chances are quite high that the mother will be impregnated by a different male this time around).

For two final examples, consider the Magellanic penguin chick that never receives any food from its parents while they cater to its larger sibling and the royal penguin embryo, still in the egg, that is kicked out of the nest by its own mother. These parental acts lower the doomed offspring's direct fitness to zero.

In these and many similar examples the fitness interests of parent and offspring are almost certainly at odds with one another. These are true evolutionary conflicts. Moreover, I presume that such fitness conflicts could be demonstrated experimentally, for instance by putting red deer mothers on a contraceptive program and assessing how much better their nursing calves do. My problem with these cases is that they are fundamentally trivial cases of parent-offspring conflict. Not a single one of them requires the explicit argument of Trivers to understand or explain it. The observed life-history (egg size) and behavioral (food delivery, pre-rut weaning) traits at issue are easily explained by the arguments sketched out by Darwin back in 1859. If these offspring are trying to use "psychological weaponry" to manipulate their parents, all are failing miserably. In each case, the parents are simply doing whatever *they* want to do, presumably maximizing their own fitness in the process. None of the subtle beauty of Trivers's insight into the asymmetries of relatedness is evident for a simple reason: offspring interests are being summarily ignored. The youngsters are too pathetically powerless.

What I would see as a really compelling case of parent-offspring conflict would have to be something that Darwin alone could not have explained and that Trivers was trying to get us to think about. Trivers's whole point was that offspring are not passive vessels into which parents pour investment, so I want to see cases where the kids win, and where in victory they gain something that adds to their fitness while detracting from that of their parents. Toddlers hollering for bubblegum in the checkout line can be viewed as *consistent with* Trivers's theory of parent-offspring conflict, but they are hardly a demonstration of it. Unless the bubblegum rots their teeth, causing them to starve to death before breeding . . . how likely are such disputes to have real consequences for fitness?

If parent-offspring conflict is to remain an important piece of theory, it needs and deserves to be tested seriously. I shall devote the rest of this chapter to some unsuccessful attempts to do this with birds, but promise to show two systems later where the case for parent-offspring conflict looks quite good.

As explained in Chapter 4, Raymond O'Connor's mortality threshold model set out just how squeezed a family's budget had to be before various members would favor one offspring's death to relieve competitive tension. The actual numbers are not of concern here, but suffice it to say that a senior sibling in a black eagle or white pelican nest would rather its nestmate be dead than fed if the death would improve its own chances of surviving by a few percentage points per day. If the junior sib's continued existence is a few percent more expensive than that, its parents should agree that the victim's death is preferable. This is when parents realize they simply cannot afford to raise two healthy young (or at least they would realize it if they had human-like cognitive skills and a bit of math background). Whatever the mechanism underlying the parents' actual behavior, which might involve some simple behavioral rules based on the body sizes or vigor of their offspring, this is where parental infanticide might appear. And recall that there is even a point farther down the road to ruin where the victim's *own* inclusive fitness could be improved if it committed suicide instead of continuing to drag down its sib.

Of special interest to us here is the indicated gap between the siblicide and parental infanticide thresholds. Picture a benign situation at first, where there is plenty of food to raise two chicks. Here the risk of being overcrowded/underfed is zero. Then imagine a gradually worsening food shortage (or likelihood of same), such that the brood's supply-demand balance is jeopardized; the danger starts inching upward, say one percentage point on Monday, an additional one on Tuesday, and so on. Within a few days the siblicide threshold has been reached and it behooves the senior sib to dispatch its nestmate, if possible—but the parents will not agree until another bad week of worsening prospects has passed. The interim can be thought of as a time window of parent-offspring conflict, the interval during which the bully sibling wants the victim dead and the parents want it preserved.

What made this offering of avian brood reduction so attractive as a model system for parent-offspring conflict was that an offspring, our one senior chick here, plays a very active role in the dispute. Far from being an unformed egg or a sullen but powerless calf, the siblicidal eagle nestling is the very agent of death, chopping away at a kinsman that parents supposedly view as its equal in value. I figured that time spent observing the post-hatching struggles in egret nests would be handsomely rewarded by offering a veritable showcase of parent-offspring conflict, with the adults serving as strict referees and preventing any serious bloodshed.

As mentioned earlier, I had a few pieces of the egret siblicide puzzle in place before the fieldwork began. First, the general pattern of brood reduction was well known for other members of the heron/egret family, with the last-hatched chick frequently being the victim.[7] Second, I had my old photograph (albeit a sample size of one) showing that great egret chicks did at least some fighting. Third, O'Connor had consolidated the relevant theory to make sense of it all. I made the confident prediction that parent egrets would oppose the early demise of the victim, perhaps by simply crouching down to brood the combatants. It would obviously be impossible to rear up tall and peck down at your rival with an adult's breast feathers in your face.

The initial field season (1979) for the siblicide study began with a series of logistical setbacks that centered mainly around my having a boat that was way too light for the seas I would encounter. My decision to base operations in Lavaca Bay, Texas, was due mainly to my having visited once a few years before, noting the large numbers of great egrets nesting on the spoil islands there, and recalling that the nests were at a convenient height (one to two meters). But that previous visit had been in midsummer, when the bay is serenely flat. Now, in the early days of April, my assistants and I were being tossed like flotsam as we ferried the sections of our plywood blind out to the colony island. Somehow, we managed to get the blind in place, but the planned observational study had to be put on hold for a while.

To salvage some information, I transported a few sets of near-term eggs back to the mainland and got them to hatch in a homemade shoebox incubator. Those chicks were moved to some nests we built on a rack beneath the floor of our stilted beach-house. We made the broods safe from local dogs, cats, raccoons, and so on by wrapping chicken wire around the captive colony. As the bay continued to make boat travel hazardous, we spent a few days tending our broods, feeding them scrap fish contributed by local shrimp boat operators.

Early on, it was obvious that the sib-fighting I had photographed years before was not rare. The captive egrets began to pummel each other as soon as they hatched, and I had to make a decision on what, if anything, to do about it. No problem, thought I, because theory predicts that the parents will not tolerate truly damaging aggression. So when the scuffling escalated to the point where bruises and lesions were becoming visible, I declared confidently that the parents would brood them. Having no parents for this task, we stuffed pillowcases with leaves and placed one on each brood with a small weight to stabilize it. The presence of this brooding "parent" had the satisfying effect of calming the tumult, though the battles resumed each time we lifted a pillow to feed the chicks.

Mercifully, I was delivered from all this silliness after a few days by a sudden calming of the seas that enabled us to extend our studies from the cage below the beach house and out into the colony itself. The honor of spending the first shift observing from the blind went to Rick Williams (the same chap who later accompanied me to Quebec). We arranged to have a conversation via walkie-talkie at nightfall, and I was eager to hear how things looked out there.

One nice thing about field biology is that reality does not hesitate to deliver periodic slaps and demonstrate that some of your most secure and cherished expectations are entirely wrong. Rick's first call was just such a slap. He reported, "The chicks are all fighting like hell out here!" So what were the parents doing? "Nothing, they're just standing there watching."

This was stunning news, but only for a moment. I turned away from the speaker and shouted, "Go downstairs and remove the pillows—let them fight!"

The rest of that summer we maintained a constant vigil in the colony, quietly recording fight after fight, with little or no intercession by the parents. I could hardly believe my eyes. How could parents just stand around while their own offspring were being killed in front of them? Scientists are supposed to be cool and philosophical at such times. We are imagined to be dispassionate and reasoned: when the facts come in completely at odds with theory, the stereotype is that the latter is jettisoned quickly. Well, it seldom works that way. The *Exxon Valdez* cannot turn on a dime, and scientists seldom forsake their predictions overnight. Part of that may be stubbornness and human frailty, but part is also caution. When expectations are not met, your eyes open wider and you take your next steps carefully, because the only thing you know for sure right then is that the picture before you is quite beyond your current comprehension.

As that first season rolled to a close, I struggled to modify my understanding to accommodate the new information. Several possibilities presented themselves. First, perhaps the parent egrets knew something I did not know; maybe they had not stepped in as feathered boxing referees because conditions were particularly bad that year. If so, and if the parents could sense that their youngest offspring was unlikely to make it, then their tolerating the battles would make sense. This was my attempt to salvage the original prediction intact: if both the siblicide threshold *and* the parental infanticide thresholds had been exceeded, there was no disagreement between parents and senior sibs. Next year, I figured, conditions might be better and whatever cue(s) parents were noticing might move them to intercede and protect the victim from its assailants.

In fact, I fantasized, I could test this conditional-intervention possibility experimentally if I could supplement the food supply for particular parents. I even began to doodle a plan to rig up some pulley ropes like those used in Lower East Side tenements for laundry, but with a plastic bucket loaded with scrap fish clipped to each line instead of clothing. In my dream, experimental pairs would receive extra food that would have to be processed by the parents (the bucket could come in, say, a half-meter above the chicks' reach). Then, if parents did predicate their interference or lack thereof on perceived food supply, those broods should be subdued by doting adults.

I had all winter to think about this plan and to entertain some others. What if parents did not stop the fights because they could not keep doing so indefinitely? The increasing food demands of nestlings reach a point after about three weeks where both parents must hunt most of the time, at which point refereeing would clearly be impossible. If parental intervention could only delay (but not thwart) siblicide, then perhaps intervention was not a viable strategy at all. Indeed, postponing the inevitable might well make the victim's death more expensive, given that it would ingest fish for three or four weeks before being killed, would be larger and harder to kill, and so on. This is a variant of the blackmailing idea, since the bully would essentially be getting its own way by making the parents' alternative even worse.

Then it dawned on me that the whole plan of waiting to see if the next season would be a "good food year" was quite unnecessary. Every season is "good" at some nests, "poor" at others, depending on the skills and fortunes of the foraging parents. While aggression had occurred at all nests, it had not been universally fatal.

The second and third seasons (facilitated by the purchase of a much larger research boat) were spent collecting observational data and getting a clearer idea of how often the sibling rivalry was lethal. We also initiated the great blue heron study for the comparative angle, as described earlier. When I looked over the full three seasons' data, we had recorded nearly three thousand sib-fights at a good number of nests, of which more than 95 percent occurred with a parent present. Even with "interference" defined as broadly as possible to include any parental act that had the effect of stopping a fight (lowering the head to regurgitate food or to adjust nest sticks, squatting to brood, even tripping and sprawling on top of the fighters), the notes showed that absolutely no interference had happened in more than 99 percent of the fights.

By then I was almost convinced. But I wondered whether parents might be doing some subtler things to extend the life of victim offspring. Maybe they do not stop fights, but slip bits of extra food to the youngest chick and keep it going for a while longer. I combed through the data and analyzed the details of the last ten meals for all chicks that had died (in comparison with same-age youngest chicks that had did not died), and again nothing emerged. Doomed chicks did not get more food (they got less), nor did they get it more easily (for example, on the rare occasions when they got a good scissor-grip, parents did not cooperate by regurgitating quickly). And once the victims were evicted from the nest and begging pitifully from the ground just a few feet below for hours on end, parents neither visited nor fed them.

Eventually I reached a full conversion. Nothing about parental behavior was consistent with the view that their fitness was being lowered by siblicide. And I realized that the whole drama is from a script written and directed by the parents, presumably because it enhances parental fitness. By stepping back from the trees and looking at the whole forest, I came to rephrase the problem less in terms of short-term behavioral adjustments that might be made after the nestlings had already hatched and were having at one another, and more on the life-history scale addressed earlier in this book. In doing so, I moved from one of David

Great egret chicks fight vigorously while their parent ignores the battle and actually yawns. Parental indifference to siblicide was a surprising finding, but is proving to be the rule and not the exception. (Photo: Douglas Mock.)

Lack's questions (Why do avian parents of many species hatch their eggs asynchronously?) to another (Why do parents produce the number of offspring that they do?), seeing the former more clearly as a restricted consequence of the latter.[8] In the process, the relationship between egret parents and their senior offspring flew away from the realm of parent-offspring conflict and into a fairly collaborative venture. Perhaps parents do not interfere with the fights because fighting and siblicide are very much in the parents' own interests.[9] This possibility, in turn, suggested the need to explore the problem of hatching asynchrony more closely.

In the fourth Texas season (1982) Bonnie Ploger joined me at the Lavaca Bay field site and we decided to perform an experimental test of the asynchrony issue. An asynchrony experiment had been growing in the back of my mind, largely from the influence of two then-recent papers. Caldwell Hahn had performed a chick-swapping manipulation of hatching patterns in laughing gulls and had obtained exactly the kind of results predicted by Lack's explanation. She made up 13 three-chick broods of nestlings that had all hatched on the same day, basically simulating what parents would produce if they simply waited till the last egg was laid before applying their own body heat and letting embryonic development begin. She compared the fates of these chicks with those recorded for the members of 48 unmanipulated control broods (where nestmates were 2–4 days separated in age). Her results were dramatically clear: the hatching pattern produced by the parents outperformed the experimental alternative by a substantial 30 percent (2.1 versus 1.6 chicks surviving to the age of 30 days). In particular, whereas all 3 chicks survived in most of her control broods, only 4 of the 13 synchronized broods remained intact. She proposed that parental rigging of unequal offspring competitive abilities might have had the effect of reducing the costs associated with sibling rivalry.[10]

A few years earlier, David Werschkul had tried a similar experiment on another egret cousin, the little blue heron, and had produced very different results and interpretation. When he swapped chicks between nests so as to make size-matched broods, he found that these synchronized chicks grew *more* rapidly than those in control (hierarchical) broods. From this he concluded that asynchrony must be a deleterious feature, wherein parents abdicate control over how food is allocated, leading to a maladaptive skewing of resources into the A- and B-chicks and unduly jeopardizing the C- or D-chicks' lives.[11] But Werschkul had not maintained a close vigil over his broods, thus might have missed important details. Like many field ecologists, he had performed his experimental manipulations and then departed, returning to the colony only for brief nest-checks every few days to measure growth and survival. This is not to condemn him for laziness: he was working alone and needed to be several places at once. It seemed to me that his most anomalous result—that all chicks in synchronized broods grew at the speed of A-chicks in normally asynchronous broods, and maybe even a bit faster—might be accounted for differently. For example, one might obtain a spurious effect of accelerated chick growth if parents had, for some reason, brought home more groceries to the synchronized broods. Because Werschkul had not observed and recorded how much food was delivered or how it was divided among nestmates, he was forced to guess at what had happened between his colony visits from the chicks' body weights alone. His conclusion, that parents of typical broods must be "unable to distribute food evenly among nestlings," was logically consistent with the data he had, but it was not the only possibility. If Werschkul was right, then this seemed to constitute a parent-offspring conflict in which the "offspring-wins" solution merited serious consideration. And yet, if hatching asynchrony is something that parents control, then why should they create a situation that lowers their own fitness? Something seemed amiss.

Bonnie and I were in a good position to fill in these details, as we had a field team of five trained observers (counting ourselves), blinds from which to watch the action, and several weeks to have a close look. Because this project began in mid-season, by Texas chronology, we could not use the great egrets that the team and I had been studying. But we had noticed that the later-nesting cattle egrets also had combative nestlings. As it turned out, cattle egrets are very similar in most respects to great egrets, only a good bit smaller (adults are about two-thirds of the great egret's height) and much more common. By virtue of their grasshopper-based diet, the boluses that parents regurgitate for chicks are discrete and small enough to be caught directly. And the beatings plus food-deprivation lead to the victim's demise in one-third to one-half of the broods.[12]

Meanwhile, half a world away in Japan and with no knowledge of our activities, a Ph.D. student named Masahiro Fujioka was putting into motion virtually the same experiment we were doing, also using cattle egrets. The two projects differed in some minor ways (for example, we adjusted the degree of asynchrony in both directions, creating synchronous broods and double-interval asynchronous broods to see what happens when the hierarchy is made steeper, while Fujioka took the extra step of replicating Werschkul's growth measures), but the coincidence was nevertheless remarkable.

Those Japanese cattle egrets did show Werschkul's odd growth result—synchronous broods gained weight as rapidly as the A-chicks in normal broods—but this time the reasons were exposed. Synchronized broods begged more intensely and somehow got their parents to hustle harder. During the first month after hatching, parents of these broods loafed less and provided more food to their chicks.[13] In Texas we also found that parents of synchronized broods brought more food to the nest: about 30 percent more than what normal broods received.[14] What seemed to be happening in broods artificially deprived of their hierarchical nature was that a dominance relationship did develop eventually, but it was much harder to hammer out and, in the interim, everybody begged at full strength. In Texas the synchronous broods fought three times as often as controls (notice once again that abundant food does not guarantee peace), with bout after bout gradually bringing the combatants into an understanding of who was dominant over whom. By contrast, in the double-asynchrony broods, fighting actually decreased.

The heightened aggressiveness of size-matched rivals makes sense. In the theoretical literature, the formal argument is called the "logic of asymmetrical contests,"[15] but it is basically the same thing that boxing promoters do for economic reasons. The business of prizefighting is all about getting paying spectators to watch the blood sport (whether live or on pay-per-view) and to wager on the outcome. If two fighters are evenly matched, or hyped as such, much money changes hands and the promoter, among others, takes a cut. Conversely, an obvious mismatch arouses less ghoulish interest: the fight will not last long and offers little suspense. Parents, one presumes, take a different view from boxing promoters. If the goal is to have a sibship that can correct itself numerically to the actual carrying capacity of the parental budget, and do so at the lowest cost, then there is likely to be some intermediate degree of unfairness that facilitates brood reduction without sacrificing chicks automatically and perhaps unnecessarily.

Note, also, that the egret parents' ability to bring home 30 percent more groceries than they normally do reveals that they ordinarily are not working at top capacity. Just as the experiment in which we kidnapped C-chicks and later restored them to the nest (see Chapter 8) showed that parents keep more food for themselves when brood size dips, this synchrony result showed that parents could be working harder than they typically do.

In the double-asynchrony nests, C-chicks were eliminated very quickly, perhaps *too* quickly, and fewer chicks survived the first month as a result. Chick mortality was also a bit elevated in the synchronous nests, but interpreting the payoff to those parents was complicated by the extra food the parents provided. We decided to divide units of success (chicks surviving) by units of effort (amount of food delivered) to derive an index of parental efficiency. If natural selection favors traits that maximize the ease with which parental effort is converted into reproductive success, as seems reasonable, then this measure should reveal which hatching pattern is best for the adults. By this measure, we found Texas cattle egrets allowed to raise broods with the normal (parentally conferred) hatching intervals to be 45 percent more efficient than those with synchronized broods and 26 percent ahead of those with exaggerated asynchrony.

The behavior of experimentally synchronized broods has now been scrutinized for blue-footed boobies in Mexico, with strikingly similar results.[16] Size-matched broods fought 60 percent more, received far more food (50 percent), and produced no more fledglings than controls, which worked out as roughly 19 percent lower efficiency for parents. Indeed, every synchronization study performed so far that has included careful quantification of parental effort has produced comparable findings: parents work 30–50 percent harder for artificially matched offspring in herring gulls, jackdaws, yellow warblers, American kestrels, and ospreys.[17]

Some remarkable evidence for American kestrels in Canada indicates that the parents of this small falcon actively *adjust* their degree of hatching asynchrony, seemingly in accordance with the anticipated family budget. In the far north, kestrels prey heavily on rodents whose populations cycle up and down dramatically. Karen Wiebe and Gary Bortolotti of the University of Saskatchewan noted that kestrel nestmates generally had wider age differences when rodents were scarce and were closer in age when rodents were plentiful. Furthermore, pairs that were on relatively food-poor territories (as revealed by rodent trapping data) and/or whose females were in below-average physical condition tended to have greater hatching asynchrony than their wealthier counterparts.

To test the possibility that parents control asynchrony strategically, Wiebe and Bortolotti gave extra food to certain pairs while not feeding others. In case you want to know how one gives handouts to hawks, Wiebe and Bortolotti simply placed two dead mice on the roof of the wooden nest box each morning for the two weeks prior to laying. The experimental and control females laid similar numbers of eggs (five), but the better-fed mothers built slight larger eggs and consistently hatched them more synchronously to have more evenly size-matched nestlings. However, these "tricked" parents (whose additional rations did not continue through fledging) suffered a drop in fledging success of about 20 percent. Using a different set of nests, Wiebe and Bortolotti also performed the now-familiar chick-swapping experiment. As with egrets and boobies, parents of synchronized broods delivered more food (about 30 percent) but produced lighter fledglings. The index of parental efficiency showed parents of normal, asynchronous broods to have a 26–56 percent advantage over those with artificially synchronized broods.[18]

To this point, I have touched on three general categories of things that at least some parent birds do that affect the slope of the sibling playing field. The laying interval and the timing of early incubation can be adjusted to produce either greater differences in age and size or nearly perfect hatching synchrony. Because avian eggs are physically separated from the mother's body soon after fertilization, they are potentially quite independent units that can be treated nicely or not. In ostriches, for example, several females lay in a shared ground nest and the dominant one (the "major hen") rolls some of the subordinates' eggs out of the nest so they die on the periphery, thereby increasing the proportion of the major hen's own eggs that survive.[19] Something quite similar occurs in the communal nests of anis and guiras, a New World branch of the cuckoo family. Unlike Old World cuckoos, these build their own nests and raise their own families. Indeed, as with ostriches, multiple females lay in the same nest and some eject the eggs of others. Here, the dominant female tosses eggs over the side, a practice that tends to delay her own laying until subordinate females have already lost some eggs. This has the effect of overrepresentation for her eggs in the eventual clutch, which is incubated by many group members.[20] In some grebes, moorhens, and pelicans, where each female lays her eggs alone, there are a few oddities like parents that lead the senior chicks away from the nest before all the junior siblings have hatched, thereby dooming the unhatched ones.[21] In other birds, some mothers adjust egg sizes as a function of laying order,[22] culminating in the remarkable crested penguin, whose smaller egg is only 15–42 percent as big as its nestmate, but more commonly by a modest 5 percent or less of volume.[23]

And after chicks hatch, parents may employ various forms of behavioral favoritism. These can be as overt as desertion or even active execution of selected progeny. American coots sometimes grab one of their own chicks and shake it to death, a practice that carries the deceptively gentle name "tousling." White storks, those picturesque monarchs that nest on European cottage roofs and chimneys and figure prominently in human birth mythology, also kill individual chicks in their own nests,[24] an act originally called "kronism" after the mythical Titan Kronos who killed and ate his own babies.

The general term "scramble competition" seems ideally suited to the way jackass penguin parents feed their offspring by leading the two chicks on a merry chase across the landscape. The jackass penguin adult approaches its two offspring then turns and runs, allowing the A-chick's superior mobility to separate them for unequal feeding.[25] And we have already encountered less aggressive types of parental partiality: food distributions that are either skewed entirely away from a designated victim or carefully delivered to a runt. Clearly there is great diversity in the way parents treat their young and a broad range of circumstances under which different offspring are likely to be valued differently.

O'Connor's prediction that parents would actively intervene on behalf of siblicide victims, though not supported by the data from the most closely studied systems, remains popular. And in fact adult interference has been reported anecdotally for Antarctic skuas (where parents may even emit phony alarm calls to interrupt fights),[26] bald eagles,[27] and black kites,[28] but the sample sizes are still much too small to be convincing. Field workers who have been continuing to study the skua have told me informally that the finding is real, so I am looking forward to seeing strong documentation. Meanwhile, a field experiment with boobies has been interpreted as providing evidence of subtle parental suppression of violence (a nest shape not conducive to siblicide),[29] but this seems far from clear-cut. It is possible that my own odyssey with the egrets has made me a bit too cautious and too prone to look at the problem from the other side. But a robust experimental study showing clear fitness consequences of siblicide (and the inhibition thereof) for both parents and offspring would be simultaneously unsurprising and exciting. There is no reason why O'Connor's original prediction might not yet lead to clear evidence for parent-offspring conflict over siblicide. And there is no reason why different species could not have evolved very different styles.

12

Parent-Offspring Conflict Revisited

Oh, I can easily imagine the fetal mishap: we were inside the womb together dum-de-dum when Leah suddenly turned and declared, Adah you are just too slow. I am taking all the nourishment here and going on ahead. She grew strong as I grew weak . . . And so it came to pass, in the Eden of our mother's womb, I was cannibalized by my sister.

—Barbara Kingsolver

Before chronicling my frustrations and failures in seeking parent-offspring conflict in siblicidal birds, I promised two cases in which I think Robert Trivers's elegant theory of such conflict is essential for explaining real biological phenomena. This chapter will deliver on that promise by showcasing those stories. But first a bit of background is needed.

Recall that siblicide seemed a likely place to look for intergenerational conflict because the senior offspring is physically empowered, thus potentially able to advance its own interests actively against those of its parents. An even better system for study, therefore, would involve parents that totally abdicate day-to-day control of the family budget, allowing the youngsters to run the nursery. Are there any forms of life in which parents simply hand over the reins to nursery residents and let *them* make key decisions—including some that raise or lower the parents' inclusive fitness—with little or no further supervision? The answer is a resounding yes. Indeed, one of the most successful of all animal lineages, the immense order Hymenoptera, offers a fantastic diversity of such families.

The order Hymenoptera embraces the ants, bees, and wasps, and there is something very special genetically about the way all of these millions of species make offspring. As new eggs ease down the mother's oviduct, they file past a balloonlike side pocket where sperm are stored.

This cul-de-sac, called the "spermatheca," contains ejaculate from one or more males. The sperm are actively motile, but they are trapped in this bag and cannot leave it unless the female's voluntary muscular system grants them the opportunity to escape and greet passing eggs. As I understand it, there is something like a kink in a garden hose blocking the sperm exit and only a contraction by a particular female muscle can unkink the hose.

This mechanism gives the mother exclusive and total control over the sex composition of her brood. In hymenoptera, the distinction between male and female offspring is not determined by the chromosomal contribution from the sperm (as in mammals, for example), but by whether sperm participate at all. Unfertilized eggs become sons; fertilized eggs, daughters. Thus, if that spermathecal muscle allows the hose to straighten and sperm to emerge, the consequent fertilization events lead to the production of daughters; conversely, if sperm are left in the spermatheca, the eggs are not fertilized and all become sons. All male hymenoptera are haploid (possessing only one chromosome from each pair of matched homologues), while all the females are diploid (having both members of every chromosome pair). The whole genetic system of these species is said to be "haplodiploid," a point whose significance will soon become apparent.

Many hymenopteran species have evolved extraordinarily complex societies, featuring great specialization and division of labor. Take the familiar honeybee, for example. A dispersing queen starts a new colony (hive), accompanied by a swarm of her sisters recruited from the old colony. She mates with one or more males and then goes indoors for the rest of her life, where her body transforms from a small flying insect into a passive egg-making machine, initially supported entirely by her sisters and then by her daughters and granddaughters. A new worker begins life in a diploid egg that is placed in a small developmental chamber where she hatches into a larva. Then she is cared for (fed, groomed, kept warm) by other workers during development. Because of her diet and other aspects of her upbringing, her sex organs do not mature fully and she is incapable of mating. She is, basically, a sterile slave whose complex job it is to raise siblings for her mother the queen, earn income for the hive, and so on.

The young worker spends her first few weeks tending the hive, not only babysitting but also helping with temperature control (by beating their wings very rapidly indoors she and her fellow workers generate and distribute heat) and actively defending the hive against threats. One particularly impressive form of defense is made by the Japanese honeybee, which must deal with marauding hornets that attack beehives in packs, like six-legged wolves. Each hornet is huge compared with an individual bee, but hundreds of the bee workers pour out of the hive and suicidally swarm onto the hornets. One hornet can slice and dice up to forty bees a minute with its mandibles, so the early going is very costly to the bees. But as the carnage continues, the ball of defenders raises the temperature of the scuffling mass to 47 degrees Centigrade, which happens to lie between the maximum heat hornets can stand (44–46 degrees) and what the bees themselves can tolerate (48–50).[1] These bees are clearly driving their own temperatures well above normal (35 degrees) via this group effort. Soon the hornet is cooked; the hive, saved.

Many of us know that a worker honeybee usually dies when she stings a human. This is because the stinger has rear-directed spines, like the tip of a fishhook, that catch in the flesh of a mammalian target. She cannot fly or be brushed away without leaving her stinger behind, which eviscerates her while also leaving behind her venom gland and a pump to do its posthumous work.

After about three weeks of hive duty, a typical worker ventures outside and becomes a pollen/nectar collector. For this, her brain undergoes a substantial reorganization, mainly the enlargement of certain structures called "mushroom bodies" needed for sophisticated spatial memory,[2] which she will need to remember the precise locations of flower patches she finds in the habitat and then to direct her fellow workers back to help harvest them. (The queen bee has a much simpler brain, sex presumably requiring less in the way of smarts.) The centerpiece of the honeybee story is their famous dance language, by which a successful scout communicates direction and distance to her newfound food source to her fellow workers. That information is coded in the angle and vigor of her "waggle-dance" on the vertical wall of the hive.

From top to bottom, then, the whole picture of honeybee sociality sounds almost like a utopian dream of cooperation and altruistic self-sacrifice for the greater cause of colony survival. Suicidal defense. Individuals laboring to raise the offspring of others. But why do they do all this? For that matter, how can a trait like *sterility,* of all things, possibly evolve via natural selection? The heart of that evolutionary process is supposed to be a reproductive race, so how can *inability to breed* ever spread through a population?

If you find this last question puzzling, imagine the aggravation it visited on Charles Darwin, who was painfully aware of the sterile castes in social hymenoptera. He referred to it as "one special difficulty, which at first appeared to me insuperable and actually fatal to my whole theory." Then he nailed it. Darwin did not even know about genes (which is a sad irony, since Mendel had already published his revolutionary work on garden peas), so he could not have had much of a premonition about coefficients of relatedness and inclusive fitness theory. Nevertheless, he surmised that there must be some situations in which selection operated *at the level of the whole family.* With that insight, the riddle of sacrificing oneself for the welfare of the queen makes more sense, since the individual worker already has virtually no opportunity for personal reproductive success, so its main route toward fitness is through its kin. Now we would say that the worker stands to gain through the indirect component of her inclusive fitness because the queen is providing her with many siblings.

It is worth noting here that the developmental reason workers have imperfect gonads is traceable to their early environment. If you lift a larva from a tiny worker cell of the hive and place her in one of the capacious royal cells, she will be fed a much richer diet (royal jelly) and grow up capable of reproduction. Every female has the genetic potential to become a queen, but worker bees carefully manage the interaction between genes and environment so as to push most into a developmental trajectory that results in effective sterility. Periodically, though, a batch of reproductives is produced. Then the larger royal cells are used and the better food is trotted out for those sisters.

So we can think of honeybee reproduction as an intermittently dual-track operation that creates two different types of broods, sterile female workers (most of the time) and the occasional reproductive brood containing both sexes. The huge production of sterile workers has often been likened to the generation of body cells in a multicellular organism:

they all cooperate for the common good, which for honeybees means the good of the hive. As this entity grows and successfully garners resources (pollen and the stabilized form of nectar we know as honey), it reaches a more classically reproductive phase in its "life." But one can easily get too carried away with such analogies, so I shall return to the firmament of gene replication.

In addition to the normal workers' inability to have a proper sex life, the peculiar genetics of haplodiploidy sometimes offers them special incentives for kin-related altruism (nepotism). To appreciate this, we need to recall the basics of sexual reproduction and the calculation of relatedness. Imagine the simplest sexual history for a queen: she reaches sexual maturity, copulates with a single male partner, and uses his ejaculate exclusively to fertilize thousands of eggs over months or years, producing waves of sterile daughters plus occasional pulses of reproductive daughters. She also produces batches of haploid sons in those reproductive cohorts by keeping her spermathecal hose kinked. Now we draw one of these sons at random from the colony and identify a rare mutant allele in his DNA. As in Chapter 2, this allele can be tagged with a prime sign, g', to show that it belongs to the general locus g but is a special and recognizable version of that gene. Next we take a daughter at random from the same hive. It does not matter whether she is a proto-queen or a regular worker for our calculation because they are genetically equivalent. The coefficient of relatedness between the male bee in your right hand and his sister in your left is the probability (a value between 0 and 1.0) that our designated rare mutant allele g' exists as an identical copy in the sister's body.

It can easily be shown from first principles that this probability is .50, just as in a diploid species, but now for a slightly different reason. If the son has the allele then we know it arrived in his body via the maternal route, that is, inside the ovum, for the simple reason that he is haploid (from an unfertilized egg): he has no father, hence no paternal genes. If he has it, his mom *must* also have it. So we know that the queen is a carrier of g', and the chance that she passes it on to any given daughter is .50 (the meiotic coin-flip). Summing the maternal (.50) and paternal (0) routes, we get .50.

In any of our more familiar diploid species, reversing the argument makes no difference, because the relatedness between a brother and sister is due to evenly matched contributions from the maternal and paternal sides (.25 + .25). But if you try it now for these two bees, something odd happens. If we suppose that the daughter has g', the chance that her brother has it is only .25. In a real sense, she is twice as related to him as he is to her. He has no chance of matching the allele if she received it from her father, and the fifty-fifty chance that she got it from her mother the queen is halved once again by meiosis.

And now for the relatedness between two full sisters chosen at random. Here too the maternal contribution is .25, leaving us to consider the paternal side of things. The chance that the known-carrier daughter got g' from her father is .50, but if he is a carrier there is no meiotic coin-flip: his likelihood of passing it along is 1.00. Remember that meiosis is a special form of cell division wherein the homologous pairs of chromosomes separate, the two members of each pair going off in opposite directions to make new haploid cells that will soon become the haploid gametes. The original parent cell is said to have undergone "reduction division," which refers to the drop from diploid condition to haploid. But when a father's whole body is already haploid (each and every cell), true meiosis with reduction division is not possible and sperm production is achieved by plain old mitosis (wherein the existing single chromosomes are faithfully duplicated before the two new haploid cells separate). And this means that every sperm is genetically identical to every other sperm as well as to every other cell in the male's whole body. And it is the identical nature of a given dad's sperm that guarantees delivery of a rare paternal mutation (or any other allele he carries) to all of his daughters. Thus the chance that the first daughter got g' from her father is .50, and if she did, the chance that he passed it along to the second daughter is a certainty. Adding the maternal ($.5 \times .5 = .25$) and paternal ($.5 \times 1.0 = .5$) routes, we get a total of .75. So, in terms of their genetic relatedness, two hymenopteran full sisters are midway between regular diploid sisters and identical twins.

One of the first major implications of this realization is that a sterile worker bee can be more closely related to a larval sister she is caring for ($r = .75$) than she would be to a daughter she could have produced by breeding ($r = .50$). This is the genetic gimmick of haplodiploidy. It provides a potential boost in relatedness that can make investment in sisters more evolutionarily profitable than investment in offspring.

Now I must confess to having made things a bit too simple. The haplodiploid story told thus far stipulated from the outset that *only one* male partner mates with the queen. And because the inflated coefficient of relatedness between hymenopteran sisters derives directly from the identical nature of that one male's sperm, causing the paternal contribution to carry its maximum value, things change dramatically if the assumption of queenly monogamy is relaxed. To see this, let us now imagine that the queen mates with two males and that their ejaculates mix thoroughly in the spermatheca while the sperm are all swimming around. The queen liberates some of these sperm, fertilizing a batch of daughters, and we draw two of these diploid offspring at random. What is the probability now that a rare mutant allele in one daughter, our friend g', will be found as an identical copy in the other?

The maternal side of things is unchanged. There's a .50 chance that the first daughter had received her marked allele from her mom and the same chance that a duplicate copy was passed along to her sister, for a product of .25. Over on the paternal side, there is a .50 chance that the first daughter got it from her father, a 1.0 chance that all his other sperm are carriers as well, but now only a .50 chance that her father is the same guy as her sister's father. That new uncertainty knocks the paternal contribution from its lofty perch, and it becomes $.50 \times 1.0 \times .5 = .25$. To say this another way, half the time that second daughter is going to be only a maternal half-sister of the first one (making their overall $r = .25 + 0$) and half the time she'll be a full sister ($r = .25 + .50$), so the average of these two probabilities is midway in between, .50.

Add a third male partner copulating with the queen, and sister-sister relatedness falls further to .417; a fourth partner takes it to .375, and so on.[3] This descent decelerates with additional mates, and the sisters' relatedness never falls below .25 (as the number of males approaches infinity, the paternal contribution creeps toward 0). So if three or more male partners inseminate the queen, a worker could gain more from investing in her own offspring ($r = .50$) than in her mother's offspring. In honeybees, this is precisely what happens: the queen accepts ejaculate from a half-dozen or so males, and it really could be more valuable for workers to invest in offspring of their own. From this we can deduce that there must be other forces shaping the evolution of the high degrees of cooperation observed in beehives. The kin-selected boost that haplodiploidy alone might have provided is not the answer, at least for honeybees. In many other hymenoptera with very different sorts of community organizations, relatedness can be much lower than .25 (since there can be multiple queens in addition to multiple males), yet some of these are still highly social and cooperative. Thus there must be some other carrots and/or sticks shaping extravagant sociality.

I will take a break from honeybees to ask whether the conditions are ever met in other hymenoptera: is the haplodiploid feature ever accompanied by the necessary conditions to keep *r* higher than .50, which could easily boost the evolution of nepotism? And are there other processes that could give a similar lift to *r*? The answer is yes to both questions. Taking the second question first, various degrees of inbreeding can push relatedness to high levels. Ordinarily we tend to think of inbreeding as "bad," associated as it can be with lineages of dim-witted historical monarchs. Inbreeding depression results when kin-kin mating brings rare recessive alleles together. Many such potentially harmful alleles seldom get the chance to cause trouble because they are shielded and thus preserved by a dominant allelic partner. The costs of dangerous recessives pairing up and expressing themselves generate selection pressures for various traits that promote outbreeding. For example, preferences in mate choice (avoiding your close kin when seeking sexual opportunities) or dispersal (all members of one sex leave the neighborhood) may have evolved in part to avoid the penalties of inbreeding depression. And, of course, such a behavioral mechanism does not require that the participants understand why outbreeding is more profitable: genes promoting such habits can simply fare better in the long run.

But there are other ways of dealing with these genetic problems, and quite a few species do substantial amounts of inbreeding, presumably because of the underpublicized benefits it can produce. A degree of inbreeding may be adaptive if the good thus achieved (keeping favorable gene combinations together) outweighs the costs of inbreeding depression.

The first question, whether *r* ever actually ends up above the parent-offspring baseline of .50, is no longer an exercise of abstract philosophical or even mathematical possibility. The tools of molecular genetics have brought it into the relatively straightforward reach of empiricism. One can simply nab a rafter-clinging colony of paper wasps and DNA fingerprint its members. From this, one can derive the average relatedness for all combinations of residents. Joan Strassmann and David Queller of Rice University did this for an array of 14 primitively social wasp species, sampling 8 to 46 colonies per species. Some of these wasps showed coefficients of relatedness that were right up in the .75 ballpark, but most averaged around .50, and only a few were down near .25.[4] It thus appears very likely that the indirect fitness dividend that derives from investing in close kin is a frequent contributor toward (but no guarantor of) heightened sociality. The boosts to *r* provided by haplodiploidy and some inbreeding presumably are important reasons why the very highest form of social organization (called "eusociality" and characterized by several traits, including some members being more or less sterile and working as helpers to the reproductives, overlapping generations, and so on) has evolved eleven times independently in order Hymenoptera and once again in another group of haplodiploid insects (thrips), but only twice (termites and naked mole rats) in all diploid animals.

Getting back to honeybees, there is one thing regular workers can do to generate personal or direct fitness. Up to this point, I have been labeling them *essentially* or *virtually* sterile, because they cannot mate and utilize sperm, but in fact they are quite up to the task of laying haploid (male) eggs. To understand what they can potentially gain from doing that rather than accepting the usual caretaker role, we must assess the two alternative payoffs to their inclusive fitness. That is, we must compare a worker's relatedness to a haploid egg of her own creation with her relatedness to such an egg laid by her mother the queen. Clearly, a self-laid male egg (a son) contains .50 of the worker's own alleles and a queen-laid male egg (a brother) has only half that proportion. With the worker's potential fitness reward potentially being twice as high if she cheats, one might expect workers to be laying eggs at a great rate.

Observation suggests otherwise: few worker-laid eggs actually exist in colonies. This puzzle has led to various attempts at explaining their absence, most postulating that workers must be under some kind of chemical control by the ruling queen. It is, after all, better from the queen's perspective to produce all the male eggs herself (each being related to her by .50) than to have some of her daughters tossing their own eggs into the brood chamber. The relatedness of the queen to such grandsons would be diluted by the intervening daughter generation, thus halved ($r = .50 \times .50 = .25$). Given this incentive, the queen was assumed to broadcast a chemical signal, a pheromone, making it clear that she is still very much in charge and insists on being the only female breeder. The problem with that argument is that such a signal is unlikely to be evolutionarily stable. If a genetic mutation arose in a worker that blocked sensitivity to that particular chemical signal, it could spread rapidly via the proliferation of worker-produced males.

So why doesn't that happen? The queen honeybee is lying in a special part of the hive cranking out eggs, which are being carried by workers to the brood chambers. She may be emitting chemical signals, but what prevents a mutinous worker from eating the egg she is assigned to carry, laying a replacement, and hauling that one to the nursery instead?

It turns out that some key sanctions are provided by the worker's sisters, her fellow workers. When I compared the interests of the queen to those of a disloyal worker, I neglected to work out how all the *other* workers ought to view the contemplated worker-laid egg. Imagine a colony where the queen has mated with exactly three males (making workers' incentive for laying their own eggs slightly higher than their incentive for caring for a queen-laid male egg). From the perspective of any worker in the colony other than the one laying the egg, such an act would be treasonous. A worker coming across a male egg need only be able to detect (presumably by scent) whether it was produced by the queen or not. If it is not the queen's egg, and if the worker has not just laid it herself, she should disapprove of its existence. That worker is related to the egg's mother (which I shall generously assume is either a full or half-sister), as we have seen, by an average r of .417 and, by extension, to the egg by half that much, $r = .2085$. And if the queen mates with more males, relatedness to such nephew eggs is even lower. Of course, the worker does not have to make any of these mathematical calculations, she needs only one simple rule: "If this egg is not Mom's (or mine), I eat it."

This means that the sin of worker laying is punished by the diffuse and individually selfish actions of siblings, a system called "worker policing."[5] It is believed to be the main reason that very, very few sons of worker bees survive.[6] And if the initial temptation to lay such an egg is evaluated in terms of its probable payoff (that son's chance of actually becoming a successful, reproductive adult), such policing sharply decreases the incentive for creating it in the first place. Why cheat if you cannot evade detection and reprisal? Thus this system is not much of a candidate for parent-offspring conflict for the simple reason that the queen's interests seem to be fully supported by the majority of her offspring, which provide on-the-ground enforcement of the royal decree.

We come at last to one of my favorite stories in all of science, which appeals to me in part because it borders on the very edges of the imagination and in part because it shows once again that real science often reverses direction. Indeed, one marvelous thing about science is that the natural world consistently turns out to be even stranger than all previous discoveries had indicated. It is kaleidoscopic and thus highly entertaining. This tale is about parasitoids, those wasps (plus a few flies) with the unsavory habit of laying eggs in or on the bodies of larger host insects, often immature beings like caterpillars, whose tissues are later consumed by the parasitoid's larval offspring. Immobilized by the mother wasp's venom, the host cannot use behavioral tricks to resist its grisly fate, but it sometimes can wage physiological war on the invaders, using immunological defenses.

There is a wide diversity of parasitoid wasps, hence a great many variations on this theme.[7] The system I want to describe concerns the family Encyrtidae, a group that practices "polyembryony." Here, an egg placed inside the caterpillar host divides repeatedly into genetically identical clones before the expanded brood's members develop into a cadre of separate bodies. The wasp mother lays only one or two eggs, yet hundreds of offspring emerge from a single cycle. Despite being ex-act genetic copies, a smallish subset of these developing larvae sometimes show a very different pattern of growth, simultaneously much more rapid and physically very different from the rest of the sibship. These "precocious" larvae have good digestive, nervous, and muscular systems, which enable them to be highly mobile within the caterpillar host. They also have oversized and menacing mandibles. On the down side, they have no circulatory, respiratory, or excretory systems, so it is obvious that they cannot live very long. They also have no gonads. In the time-honored tradition of assuming that everything not immediately understood must be some kind of freak, they were originally assumed to be accidents of development, curious "monstrosities."

Further research indicated, on the contrary, that the precocious larvae probably have great value to their genetically identical siblings by serving as bodyguards (hence the jaws). In experiments where the host caterpillars were invaded by competing parasitoids of three other species, Yolanda Cruz showed that the precocious larvae attacked the uninvited guests, thus securing the host tissues for their slower-growing sisters.[8]

A decade later the story changed yet again (though this version concerns a different genus from the one Cruz studied, thus may be a truly different story). In the parasitoid *Copidosoma floridanum* the mother wasp inserts either a male egg (haploid) or a female egg (diploid) or, in most instances, one egg of each sex into the egg of a cabbage moth. A flurry of polyembryony ensues, generating an impressive crowd of roughly 200 male clones and 1,200 female clones, so the mixed sibships are about 85 percent female. The emerging precocious caste, numbering 30–50 larvae, turns out to be nearly all female. The other piece of information we need here is that incest is extremely common in this species: brothers provide sperm to their own sisters and most mating occurs before anybody actually leaves the natal caterpillar. Before we get to the juiciest parts, though, I must introduce another candidate system for parent-offspring conflict.

Shortly after his theoretical paper that introduced parent-offspring conflict, Trivers and his co-worker Hope Hare published a fascinating paper offering a great angle on the problem.[9] It concerned the ideal sex ratio for reproductive broods in eusocial hymenoptera and identified an issue they called "worker-queen conflict." Throughout this chapter I have been building up the background needed to appreciate their key prediction, so here goes. The idea is that the sporadic occasions when a reproductive brood is being produced offer brief moments of great conflict of interest between the mother queen and her existing worker daughters. Being equally related to both daughters and sons ($r = .50$ to each), the queen should generally favor equal production of fertilized and unfertilized eggs. And, because she is the only one in charge of whether her spermatheca is open or closed, an equal sex ratio is generally what she creates. So far this example looks like just another parent-wins case of what I regard as trivial parent-offspring conflict: the parent is getting her own way, as usual. But the beauty of worker-queen conflict, as already noted, is that the queen may not be in any position to enforce her preference once the workers haul the eggs off toward the brood chamber. It is the workers that have ultimate control over the *fates* of male and female eggs. (As Stalin once observed and the 2000 U.S. presidential election demonstrated, it doesn't matter who gets to vote; what matters is who counts the votes.)

Do the workers care whether the sex ratio is balanced or not? Trivers and Hare argued that they should. Recall that a worker is related to her full-sister eggs by $r = .75$ and to her brother eggs by only $r = .25$. It follows that a worker's fitness, which can be measured in terms of the lifetime breeding successes of her sexually capable siblings, will fare better if the total investment is skewed three to one toward sisters. Indeed, even if the queen has had several male sexual partners, the workers' relatedness to sister eggs never falls quite as low as the .25 ceiling for brothers. Accordingly, these empowered workers ought to favor a secondary adjustment in the brood's sex ratio toward female sibs. And some preliminary data Trivers and Hare marshaled on reproductive broods of ants did seem to show female-biased skews in many species.

Let's return to those *Copidosoma floridanum* larvae growing up inside the cabbage moth caterpillar, and let's assume the mother wasp laid two eggs, one male and one female. Nearly the whole warrior caste belongs to the sisterhood, and these are not mere full hymenopteran sisters, these are *clones* ($r = 1.0$). With their massive jaws and nonexistent futures, they provide protection against invasions (from eggs inserted by a rival species),[10] and they also operate as a hit squad against

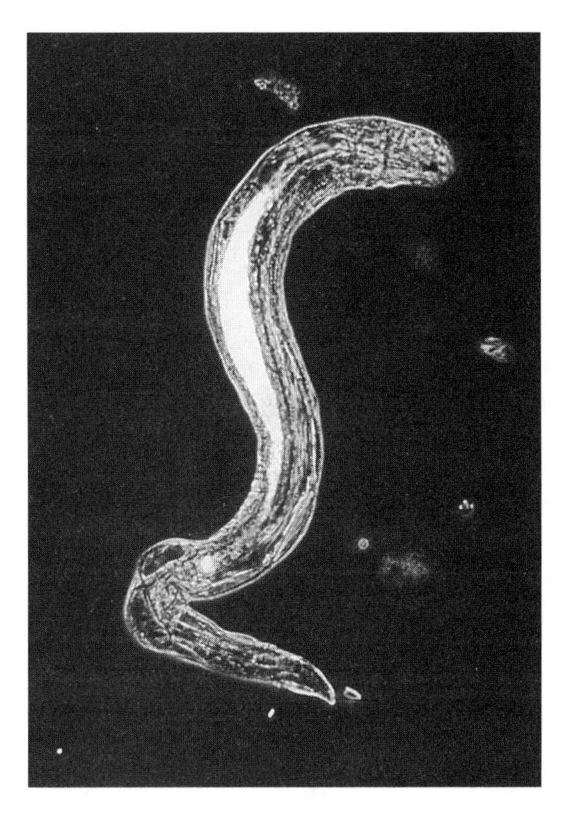

A female parasitoid wasp larva of the precocious or warrior type has a very short life expectancy and no gonads. Her job is to kill parasitoids of rival species introduced into the host caterpillar after she and her siblings have hatched and also to kill many of her own brothers. The central area within her body (pale gray in this photograph) is light-colored because of tracers that demonstrate that she has eaten her brothers. (Photo: M. Strand.)

their own brothers. That is, the precocious larvae crawl around killing and eating their male siblings, but not their sisters. The brothers quite reasonably are unenthusiastic about this activity and hide in the moth's fat deposits, but many are killed and the sex ratio swings hard toward female reproductives.

When the two eggs were laid, recall, the sex ratio was one to one, and at the end of the polyembryonic expansion it had gone to 85 percent females (not even counting the doomed precocious warriors), but the female proportion climbs even higher as the result of the sex-biased siblicide.[11] This apparently serves the interests of the warrior larvae because their siblicidal activities skew the allocation of limited caterpillar tissues into the construction of more valuable sisters, especially considering that it may take only a few brothers to inseminate all the reproductive females. Of course, the warriors probably had better not wipe out *all* the males for the obvious reason that the brothers that manage to survive are needed to inseminate the reproductive sisters. In fact, once the sisters have been serviced the surviving males develop wings and fly off for a couple of weeks of free time, presumably searching for the few pockets of unmated females that emerge from caterpillars where the mother laid no male egg in the first place.

Now we are ready for the two cases that I concede are good examples of biological features that almost certainly evolved in response to parent-offspring conflict.[12] Remember that my three key criteria for this theory are that: (1) parents and offspring appear to "disagree" about something (the best solution for parental fitness interests and that for offspring interests do not match); (2) the difference at issue is not trivially small (no candy bars in the checkout line); and (3) the offspring get their way. My third stipulation bothers many biologists, so I remind everyone that parental bullying was well explained by orthodox Darwinian explanations and is, essentially, the status quo. We seek a few convincing "man bites dog" exceptions.

The first such case concerns the anatomy of parasitoid wasp larvae much like those just discussed. Most parasitoids do not have the trick of polyembryony in their arsenal, so each mother creates the family size that works best for her (by adjusting the number of eggs placed in or on the paralyzed host). Then she leaves forever and has no further influence on the drama: she has abdicated and the offspring can adjust things to their liking. One action offspring can take to promote themselves as individuals under such circumstances is siblicide, as we have seen. In particular, if the mom lays ten eggs and the first-hatching larva prefers not to share the host with nine other hungry mouths, it can execute the rest of the eggs or emerging larvae and have the whole host to itself. For this gruesome business, killer jaws are an asset.

A survey of several dozen closely related parasitoid species (all in the same genus) revealed a fascinating relationship between the presence of massive mandibles and tiny clutch size.[13] Species in which larvae possess killing-style jaws tend to lay only one egg per host, while those whose larvae have less lethal-looking mouthparts produce larger families. This implication is that offspring capable of dispatching unwanted nurserymates have undermined the maternal preference for laying more eggs, thus forcing their own mothers to set them up as sole occupants of the resource base. We infer that the maternal optimum is for more than one egg both on the basis of logic (insurance value, and so on) and because the laying of multiple eggs is so common in the non-jawed species. In short, this looks like the end result of an offspring extortion scam that worked; the parent had no choice but to pay.

The clearest documentation for behavior having been shaped by Trivers's parent-offspring conflict comes from an experimental study of worker-queen conflict over brood sex ratio. One of the challenges in research, as we have seen, is that clear interpretation requires decisive elimination of alternative explanations for the phenomena observed. With worker-queen conflict, recall that the premise is that the mother favors a one-to-one sex ratio in the reproductive brood, while the workers prefer one biased toward daughters (up to three to one for full sisters). This becomes tricky if there are other biological features that might push the mother toward preferring more daughters.

We just learned of one such feature with the encyrtid parasitoids. If a species has brother-sister mating within a closed nursery before anybody disperses (called "local mate competition"), the mother's fitness is likely to be better served by creating a surplus of female offspring (because just a few sons can fertilize her many daughters). In fig wasps, for instance, where newly pupated brothers inseminate their sisters before anyone leaves the natal fig, mothers lay many more diploid than haploid eggs.[14] This complication, along with several others, was pointed out soon after Trivers and Hare proposed worker-queen conflict.[15] But over the years several lines of evidence have come to light that have sustained interest and confidence in the original argument based on relatedness. For example, there are a great many hymenopteran species with female-biased broods that do not have brother-sister mating, so local mate competition cannot be operating in those systems.[16]

Some of the exceptions proposed by Trivers and Hare, where brood sex ratio does not hover around three to one, have also stood up well. For example, ant colonies that have multiple queens have much lower sister-sister coefficients of relatedness. (To keep the discussion simple, I have focused on haplodiploid societies based on a single queen. With more than one queen, as in many paper wasps and other hymenopterans, *r* declines well below .25.) From the relatedness argument, such sisters have less to gain from skewing the brood's sex ratio toward females, and these species seem to have appropriately lower proportions of females in the reproductive broods. At one extreme, in the so-called slave-making ant species, many of the workers are physically abducted as eggs from other colonies (often even other species), thus are totally unrelated to the local brood. In these, the workers should have no genetic inducement to skew brood sex ratio, and in fact it is approximately one to one.

More convincing evidence concerns a key event that recurs from time to time in single-queen colonies and provides an improbable prediction. When the old queen dies and her reproductive role is inherited by one of her daughters, everybody's fitness interests should change dramatically, but only briefly, before snapping back. Here's the argument. Just before an old queen's death, the workers are expected to view the sex ratio as described above: they should favor an investment skew toward sisters. But as soon as the old queen dies one of her daughters ascends to the throne and assumes the egg-laying honors. Suddenly the preexisting workers are no longer raising siblings (with *r* values up to .75 providing the incentive for bias), but nieces and nephews. For the moment, their inclusive fitness interests are channeled through their sister, the new queen, which means that their views on offspring gender ought to concur with hers. Because she, as the reproductive, favors a one-to-one investment in sons and daughters, so should they.

Furthermore, this situation should last only one generation. When the next cohort of workers take over in service to the new queen, the offspring *they* invest in will be siblings and so their optimal investment ratio should jump back toward females. This transition has been examined in sweat bees and, as predicted, when a queen is replaced the brood sex ratio does shift away from the female bias and closer to parity, then shifts back a generation later.[17] Well and good, so far.

Ulrich Mueller performed a field experiment to determine whether it is in fact the transition from old queen to new (and not some other possible covariant of the situation) that leads to the sex ratio change. He removed the single breeding female from nests of another primitively eusocial bee, *Augochlorella striata,* and got the same effect: male representation in the subsequent brood rose temporarily.[18] This phenomenon seems most reasonably explained by the Trivers-Hare argument, and it is the best case I know that both involves asymmetry in relatedness and shows offspring getting things their way at their parents' expense.

13

Till Death Do Us Part

Mothers are fonder than fathers of their children because
they are more certain they are their own.

—Aristotle

Had England's King Henry VIII known that offspring gender in humans, and all other mammals, is determined by the presence of a Y-chromosome, which can only come from the *male* parent, the Tower of London might have seen less traffic in doomed queens and the Church of England might not have been invented to sanction his divorces. The double requirement that Henry produce an heir who was both legitimate and male created obstacles. He cleared the first of these by finding the key to wedlock, but chromosomal sex determination was not discovered until the twentieth century.

The marital relations of ordinary citizens are complex enough without the future of the realm hanging in the balance. Most Western societies expect their citizens to practice monogamy, at least publicly, but even the simplest forms of this mating system involve two adults agreeing at some level to share various burdens, responsibilities, and obligations. How the work actually gets divided between mates can be the result of labyrinthine, and often lifelong, negotiation and renegotiation. Given what we know about how social behavior evolves, this should not surprise us. At stake in most such "pair-bonds" is the reproductive success of both parties, so the fitness incentives are very high. Simultaneously, since raising offspring is an expensive process, and since the two mates are usually unrelated genetically (to avoid inbreeding depression), there is also great potential for fitness conflicts within the partnership.

I have been discussing things in human terms now for two consecutive paragraphs, surely a record for this book, but the real epicenter of monogamy is in class Aves. Birds have good reasons for practicing monogamy. First, feather-bearing offspring are expensive and difficult to raise, so single parenting is often not an attractive option: two parents working together, however imperfectly, may achieve much higher success than a soloist. Second, the fertilization of an avian egg is followed swiftly by the physical dissociation of that proto-offspring from its mother (nearly always within twenty-four hours), so neither parent is automatically trapped into a deeper commitment than the other.

This second reason contrasts with the situation in mammals: with lengthy pregnancy and ensuing lactation, female mammals make a much larger early investment in each offspring than do the relatively unencumbered sperm donors. In many birds, males can do just about everything but lay eggs. They often share fully in incubation, defense of the nest, and feeding of the chicks. To be sure, there are avian species that are not monogamous, and in them the split in parental labor ranges from 100 percent male (the mound-builder we've met, various sandpipers) to 100 percent female (most red-winged blackbird populations). But by and large, birds are great practitioners of the two-parent system. Politicians sincerely wishing to understand Traditional Family Values would be well advised to spend more time contemplating backyard songbirds.

Theoreticians like to start an analysis with a simple and even idealized form of the system under study—a stripped model—so I shall introduce an abstraction that Geoff Parker calls "True Monogamy."[1] Here, sexual exclusivity is formally required to be both absolute and lifelong, so that not even the death of one partner permits the survivor to find a replacement mate. Under such hypothetical rules, there would be no sexual conflict in the evolutionary sense because neither partner can profit (reproductively) from foisting any of its own burden onto the other. That is, each will achieve a lifetime fitness score precisely equal to that obtained by its mate, which means that the best selfish strategy is to do whatever also enhances your mate's longevity and breeding success. Your two fates march in lockstep. Of course, as soon as the stipulations of absolute and lifelong fidelity are relaxed, the interests of the two mates cease to be perfectly congruent and the potential for exploitative selfishness begins to creep in.

Before this goes any further, a fair question is whether such True Monogamy actually exists in wild birds. And the answer is no, not quite. Even though approximately 90 percent of bird species are classified as monogamous,[2] the great majority of these are more correctly called "socially monogamous" (a term that means they raise young in pairs but does not imply either the presence or the absence of mating exclusivity), and they do not necessarily even retain the same social mate over subsequent breeding cycles. A few species, however, show amazingly durable pair bonds, with the geese and swans having particularly notable reputations for fidelity.

The champion is apparently Bewick's swan, where a sample of 919 pairs showed not a single divorce (that is, not a single record of an individual breeding with a second partner while its previous partner was still alive); but even these birds re-pair if the original mate dies. Considering how long swans can live (easily a quarter of a century), partner loss is not unheard of, and one Bewick's swan has been recorded mating with seven sequential partners.[3] The closely studied barnacle goose has numbers only slightly less impressive. In a sample of over 6,000 marked birds, 99.6 percent were socially monogamous, roughly two-thirds had only one mate during their whole lifetimes, and those that took any additional partners nearly always did so because of the original mate's death (87 percent).[4] All in all, this is pretty close to True Monogamy. So one would not expect to find much sexual conflict in these particular waterfowl. If American humans were this monogamous, there would be no cheatin' and we probably would not have invented country music.

By contrast, while the European blackbirds in the Cambridge Botanic Garden are also socially monogamous, their relationships dissolve much more easily between seasons. Fewer than half of all surviving adults retain the same partner for the next nesting effort. In many cases this is because the old mate has died, but in nearly a third of all pairs that suffer no over-winter mortality both take new spouses anyway. Divorce here seems related to females being a bit more mobile than their highly territorial mates: it appears that females may simply relocate to more productive parts of the garden between seasons.[5] In blackbirds, then, each individual's future fitness is not tied to the shape its partner was in at the end of the previous nesting cycle. That is, wearing out one's mate arises as a viable option.

Great blue herons negotiate a seasonal pair bond. A male *(right)* that has been displaying in the nesting colony allows a female to approach for the first time. Both are wary, as revealed by their erect crest plumes. Courtship is a complicated process for monogamous birds, whose partnership will involve substantial investment by both parties. (Photo: Douglas Mock.)

The extreme in monogamous fluidity is shown by greater flamingos nesting in the Camargue of southern France. Again, the pairs are socially monogamous, but *everyone* takes a new partner between breeding seasons.[6] So here it would seem that concern for the partner's welfare might be at very low ebb indeed.

When monogamous mates may not share a common future, then, selection can favor each trying to provide less than its full share of costly parental investment. This can be thought of as an attempt to parasitize the partner. Such selfishness should be checked by any extrinsic factors that impel the relationship to continue, for the mate to be retained (such as an organized church, risk of being publicly stoned to death, or, more simply, a dearth of alternative mates). The amount of work one can get away with dumping on one's partner is also likely to be limited by the fact that the partner should view the situation in much the same way but from the diametrically opposite position. That is, the chief constraint on how much a male can exploit his mate may well be her desire to exploit him.

One can imagine this leading to a series of sequential negotiations, involving two "partners" of roughly comparable power and need for the other's assistance. Such coevolutionary struggles have been likened to those between union leaders and management, where one side assesses what the other will offer and predicates its counter-offer on how that fits with its own best interests.[7]

The negotiations between two caregiving parents have been modeled by theoretical biologists in various ways,[8] and one game theory version can be summarized roughly in words.[9] The essential selfish interests of, say, a courting male can be considered in terms of how much he is willing to invest in offspring relative to what his mate will supply. Lacking True Monogamy, we expect our focal male to be seeking a bit of a bargain, and we expect the same of the female he's courting. Each would like to achieve a full measure of reproductive success (healthy offspring raised to independence) while expending somewhat less than a full measure of effort, but neither has the option of turning the whole job over to the other.

The process of bidding and revising the bids is imagined to take several steps, which might mean a lengthy and complicated courtship wherein each is gathering information about the abilities and proclivities of the other. Without implying human-like calculations, each is imagined to begin with the equivalent of a low-ball offer, then to reconsider its position in light of the prospective mate's counter-proposal. We might imagine that the male offers, say, fifty units of effort and the female "replies" that she could only go to forty with such a partner, causing him to probe with a second offer of sixty, inspiring her to raise her ante to fifty, and so on. The details of the model are not important here, but the fundamental assumption that each party follows its own self-interest produces the result that there is one and only one pair of values (male bid matched with female bid) that both will accept. In this two-player game, there is a single equilibrium, which is the Evolutionarily Stable Strategy. Not surprisingly, it makes neither side happy, but both find it tolerable because it's the best either can do.

How does this help us think about shared childcare and monogamy? The answer to that question is the connection between theoreticians and empiricists. A valuable hypothesis like Hamilton's rule leads us to look for things we had not previously seen as worth exploring, but, like a compelling political cartoon, it is inevitably and intentionally an oversimplification. I like to think of theoretical models in parallel with Picasso's definition of art as "the lie that helps us see the truth" and Darwin's view that "without a hypothesis there is no observation." A good piece of theory is like a flashlight on a dark night: it can help us find our way with fewer missteps. We also must test the light itself from time to time to be sure it is not creating illusions and leading us astray. And the light shines only so far into the gloom. For example, the ESS model above suggests that parents eventually settle upon a division of labor that is a compromise between their conflicting interests. That says nothing explicitly about whether their combined effort actually meets the needs of the offspring, but it strongly suggests that each party is likely to choose a workload somewhat below the maximum of which it is capable. And if parents are holding out on their offspring (by, say, cutting back on food deliveries) for any reason at all, sibling rivalry is going to be exacerbated.

In truth, we have long known that parental workloads can be quite flexible in response to both natural and experimental fluctuations. And it is also clear that offspring signals vary in vigor (sometimes loud and frantic-looking, other times desultory). The general impression is that parental responses to signals from offspring tend to be positive: when begging is strong, parents hustle; when it is weak, parents relax.

But there has to be an upper limit to parental sensitivity, a point beyond which the adults no longer increase their efforts in response to youngsters' more insistent begging. We are a long way from understanding how that works. The dynamic between chick begging and parental effort has interested researchers for decades, but without much progress until recently. Just after World War II, Lars von Haartman made one of the first demonstrations of parent birds' sensitivity to their young by designing a nest box for great tits that contained, in addition to the regular nest chamber, a special hidden compartment where a second brood of chicks could be concealed. The resident parents thus were exposed to an amplified signal barrage while tending their own young (needless to say, this design became obsolete with the invention of portable tape recorders). These parents dramatically increased their rates of food delivery.[10]

We have already seen some evidence of "honest" signals by reed warbler chicks and the responsiveness of attending parents. And we have seen cattle egret and other parents that boost deliveries when brood members are size-matched and cut back when chicks are removed. These parents must be noticing something different about their broods when they alter their own behavior. In one European starling nest, researchers juggled brood size repeatedly, starting with the original five chicks and then shifting the number daily, with a low of two and a high of nine. Parental delivery rates tracked these fluctuating brood sizes qualitatively (highest with nine, lowest with two, and so on), but not in direct proportion to the number of hungry mouths (the per capita amounts were lowest with nine, highest with two).[11] In short, parents appear somewhat sensitive to the whole brood's combined signals, but they are not simple marionettes that dance to offspring demand.

One especially clever experimental approach to exploring predictions from the ESS negotiations model was developed in a field study of starlings in England by Jon Wright and Innes Cuthill. One member of each pair was physically encumbered with small lead fishing weights crimped onto the bases of several tail feathers. These were not especially heavy burdens, but they did alter the distribution of weight on a previously well-balanced aircraft, and they slowed food deliveries by 10–15 percent in the partner hauling lead. As the model predicted, compensation provided by the unburdened mate was only partial. That is, if the handicapped parent abridged its deliveries by 10 percent, the partner might increase by half that much but not enough to take up all the slack. It did not matter which parent was handicapped: males and females were equally willing—and equally limited—when it came to filling in.[12]

Of course, parents have a finite capacity for working on behalf of the current brood. If they work much harder than normal they may suffer the biological equivalent of job burnout, jeopardizing their future reproductive success. But there is no reason to assume that parents confronting stressful conditions necessarily take the burden on themselves. When family-raising tasks prove unacceptably expensive, parents have the option of sacrificing the offspring instead, either drastically through desertion or selective infanticide or, more passively, by simply doing an insufficient job of providing. Cutting back on effort may allow them to maintain their own condition, but it passes the misery on to the current brood.

In various fishes, for example, the male parent alone remains with the fertilized eggs and provides all their care (guarding them, fanning them with oxygenated water to discourage fungal growth, and so on). His babysitting can establish a trade-off between his own well-being and that of his offspring. Specifically, he is constrained not to leave his egg mass (many predators are ready and willing), and this greatly limits his feeding options. As a consequence, he may sustain himself by cannibalizing some of his own eggs.[13] In some cannibalistic species offspring have even been shown to be superior nutritionally to alternative foodstuffs.[14] But, of course, they are also an expensive snack in terms of Darwinian fitness. If the male's primary objective is to reproduce successfully, consuming his own offspring is not an obvious route to success.

There are other variations on this parental cannibalism theme. A male three-spine stickleback, while guarding his nest, may sneak out and steal some other male's eggs, depositing them back in his own nest. Superficially this looks idiotic, the equivalent of a reed warbler parasitizing itself by recruiting a cuckoo egg for its nest. But if the male cannibalizes the eggs in his nest, perhaps he is gathering these non-kin for his next meal. That, too, is probably not his main incentive for stealing them. It seems that such eggs help him attract additional sex partners. As you might have guessed, sticklebacks and other practitioners of this habit are not monogamous. A successful male can tend the eggs of female #1 while courting female #2, then #3, and so forth. Females are more strongly attracted to males that have many eggs in their nests. One possible reason for this is that a male with few eggs might be a bad risk, an inept guardian. Another is that a nest already containing many eggs offers a safety-in-numbers opportunity for a female's own eggs. Also, given that the male himself is an egg predator, a male with plenty of eggs already on hand may already be well fortified with potential dinners and thus unlikely to consume the new female's eggs. And, of course, there may be other reasons that have not yet occurred to us.

Returning to monogamous birds, one way to answer the question of which family members suffer from penurious parenting is to tinker with the workload and see who gains or loses. One of the nicest demonstrations of this was performed by Lars Gustafsson and his colleagues on an island population of collared flycatchers, dapper black-and-white European songbirds that nest obligingly in special boxes (as well as the natural crevices and woodpecker-excavated cavities that were its only choices in the past).[15] Gustafsson's study population is nonmigratory; in fact they never even leave their home island. This highly sedentary habit, called "philopatry," greatly facilitates the study of deferred consequences.

To assess the possible costs of parental effort, the experimenters either added one or two eggs, removed one or two eggs, or swapped eggs between pairs of control nests without any change in number (so that the control families experienced the same amount of visiting and egg-handling as the experimental families). The idea was simple: if parents normally set out to raise about six offspring, and if their task is either increased to eight or trimmed to four, then either the parents will have to adjust their workloads to fit the challenge or the welfare of the nestlings will be affected. A total of 320 flycatcher families were involved in the project. The enlarged broods showed a brief advantage, producing slightly more fledglings. But these young did not fare well *after* leaving the nest, and overall the enlarged broods produced the fewest breeding adults in the ensuing seasons. So one cost of overtaxing parents is that the chicks apparently get less food and have a dimmer chance of surviving the winter.

The survival of the parents, both male and female, was not affected by the size of family they raised, with equal proportions of adults showing up in the ensuing breeding population in the next year. But that only shows that parents do not literally work themselves to death. Scrutiny of the parents' subsequent reproductive performance revealed quite a different pattern. In particular, the mothers' reserves showed a very strong relationship with the workload they had borne. In the group from which two eggs had been removed, mothers laid larger clutches in the following year. By contrast, mothers that had tried to feed two extra mouths laid fewer eggs. Most unexpectedly, even the *daughters* from egg-reduced families laid more eggs than daughters from enlarged ones. (We know less about how the fathers and sons fared in all this because male reproductive success is a lot harder to track accurately.)

Thus it seems that modest decreases in per capita resources provided by parents, from whatever cause, can affect any or all of the social dimensions within the family. A food shortfall immediately exacerbates sibling competition, raising the likelihood that brood reduction will be necessary (or, if a species has no pressure-release mechanism such as hatching asynchrony, that the whole brood will perish). And it may also influence the nature of any parent-offspring disagreements. Superficially, an inadequate family budget would seem to foster conflict between parents and dominant siblings, by pushing the family's deprivation beyond the threshold where parents and offspring are supposed to disagree about whether a sibling should die.[16]

Some theoretical modeling by Geoff Parker, though, has revealed even subtler dynamics.[17] When food conditions are bad, an effective sibling size- or dominance-hierarchy automatically concentrates the burden on the shoulders of the lowest-ranking nestling, which then loses condition and becomes less and less valuable to everyone else in the family. As it weakens, the designated victim also becomes less *threatening* to the others. As it nears the brink of death, the victim is already so weak that it carries very little potential indirect fitness for its siblings. With so little at stake, nestmates may be indifferent about finishing the execution, so the victim may be left to languish. By the same logic, a dying victim carries virtually no potential direct fitness for its parents. In short, the decline of one handicapped-in-advance brood member, whether due to environmental factors beyond the parents' control or to behavioral choices by family members, is unlikely to end in a sudden and precipitous drop in value. This insight may help us understand why parents seem so detached during siblicide dramas: at any given moment they may have little at stake.

One of the least understood things about sibling rivalry is whether temporary periods of deprivation (say, being underfed for a few days or weeks while more senior nestmates are helping themselves to the choice meals) have long-term effects on junior siblings. The "brink of death" is unlikely to be well defined, and being "half-dead" may not be as reversible as we tend to think in this era of ambulances, emergency rooms, antibiotics, and painkillers. This topic is not an easy one to study, considering the necessity of documenting the early problem period and then finding the same individuals later as adults and recording their breeding success—particularly because many species (including the great egrets and cattle egrets I study) disperse widely after fledging, colonies form and dissolve, and so on. But I am very curious about what happens when the designated victim does not die. Can it recover fully, or is it permanently below par?

I am aware of only two studies on brood-reducing birds that compare the long-term fates of chicks from the bottom of the family hierarchy with those of their more buffered senior siblings. For a population of a species called western gulls nesting on Southeast Farallon Island just beyond the mouth of San Francisco Bay, Larry Spear and Nadav Nur analyzed 352 three-chick broods whose hatching order had been marked. As usual, C-chicks had a tougher time early on, with more than half dying without reaching the age of about three months, when western gull fledglings become fully independent (their early mortality was double that of A- and B-chicks). But the C-chicks' disadvantage vanished beyond that point. While only half of all fledglings that reached independence survived to reach the age of first breeding (about thirty-three months), the ones that had been C-chicks held their own proportionally.[18] Thus the competitive disadvantages that parents impose on their youngest offspring may be ephemeral (unless, of course, they are fatal).

There are many other factors complicating this situation, including the quality of the parents themselves. In many monogamous birds, including some gulls,[19] pairs that work well together in a given season tend to divorce less often and enjoy the benefits of their successful partnership in the future, while those whose previous arrangement led to failure tend to find new mates. Whatever the reasons, a strong parental team may well provide more food than a shaky one. In that light it is interesting that the single strongest predictor of western gulls' survival is the month of fledging: early birds really do get the worms (as well as the rotting fish, the offal, and other treats).

Other evidence of how a chick's rank within its family affects its future comes from a field study of egrets. Unfortunately, not *my* egrets. I stopped banding egret nestlings long ago because the chances of ever finding them again after they dispersed seemed low. But a more optimistic and tenacious team of biologists in southern France, led by Heinz Hafner, has shown what can be accomplished. Over a period of fifteen years, Hafner and other researchers from the Tour du Valat Biological Station banded and/or wing-tagged 3,788 nestlings of the little egret. (An observational study of the little egret in Japan suggests that it is a brood-reducing species that uses little or no overt aggression, as also seems to be the case for its closest relative in the United States, the snowy egret.)[20] The researchers kept careful records on the nestmates' "size ranks" at the time of banding. Wishing to minimize disturbance to the nesting colonies, they did not mark each chick on its day of hatching, but instead waited till the broods were about three weeks old, then approached each nest just once, quickly grabbing all the chicks and banding them with numbered and colored leg bands, using one distinctive color combination for the largest brood member (presumably the A-chick), another for the second largest (B), and so forth.

Egrets in general start breeding when two years old, and the field team combed all the local colonies season after season searching for banded adults. Whenever they spotted one of their former nestlings, a "blind" observer (one who did not know the rank this parent had had back when it was a chick) recorded the number of young it was raising 20–25 days after hatching had begun. The researchers never managed to find pairs in which *both* members happened to have been color-banded as nestlings, but they did eventually accumulate 56 records where the early rank of one parent was known.

The results were remarkably clear: adult pairs containing at least one former A-chick (with the other parent being of unknown rank) were raising larger families than those containing a B-chick parent, which in turn were larger than those of former C- or D-chicks.[21] If similar effects of nestling rank on future breeding success exist in other species, the results suggest that the temporary privations experienced early in life impart long-lasting handicaps. The jury on this matter is likely to remain out for some time because these studies are so difficult to do well. As well, the contrasting results between the western gull and the little egret suggest that many patterns may be possible.

Getting back to the division of labor in two-parent families, we may wonder how the balance between male and female effort is maintained through the caregiving period. What prevents either party from unilaterally violating the negotiated solution? Do the two adults monitor each other's workload, somehow keeping track of how much the other is contributing? Something along these lines seems an essential component of the cooperative venture. Finally, if such a scoring system is in place, what happens when one parent detects that its partner is cutting back? A drop in effort could arise for innocent reasons, such as a temporary run of bad luck while foraging, or because of a physical ailment, or because of plain old goldbricking (which we might dignify by calling it strategic mate-exploitation). It is easy to imagine that the best response could hinge on a parent's estimate of how long the partner's performance will fall short. If the problem appears to be temporary, the optimal response may be to take up the slack fully for a little while, lest the kids needlessly suffer permanent damage.

In a series of house sparrow studies that Trish Schwagmeyer and I have been pursuing since 1994, we have witnessed quite a few cases that are hard not to interpret anthropomorphically. The two parents are usually flying back and forth, each making a dozen or more trips per hour as they bring caterpillars and smaller insects to their hungry charges. Unless it is so chilly that the chicks need to be brooded, food deliveries tend to be very brief, taking perhaps five to twenty seconds. But sometimes one parent starts to loaf, delivering the food and then spending several leisurely minutes perching on the nest's roof or nearby. From that vantage, the indolent parent can see across the open habitat for considerable distances, and it usually stays put until its partner approaches with its next delivery. Then it gets moving and heads off to forage, as if it has been industrious all along.

Sparrow parents that start goofing off like this are likely to do so for several deliveries in a row. Their partners sometimes attack them, either giving a squawk and a short chase or even catching hold of their tail feathers so the two fall to the ground in a fluttering ball. That puts an end to loafing. Objectively, this may be a simple case of effective punishment in action. Indeed, the classical psychological concept of punishment has been recast simply as a means by which the net benefit for a particular activity, in this case loafing, is rendered negative by increases in the social costs one must pay for it.[22] From that perspective, it is not surprising that punishment is a common feature of animals' social behavior. (Recall the way cattle egret B-chicks use head-pecking to make begging, or even head-raising, an unaffordable activity for C-chicks.) Whether parental division of labor can be enforced generally through such penalties will be interesting to explore.

The slight to moderate adjustments in parental delivery rates discussed so far represent the subtle and refined end of a continuum of which the other extreme is total abandonment. There are, to be sure, situations in which one adult dies suddenly, and its partner faces the dilemma of what to do next. On the one hand, a newly widowed parent can try to finish raising the current offspring by itself, in which case its own future potential for reproduction may be compromised; on the other, it can give up on the current youngsters as a lost cause and seek a new partner. (There are cases in nature of single parents finding a new partner willing to help with the existing family, but they are relatively rare.)[23]

The best solution available to a newly single parent is believed to hinge on such matters as how much additional care the current youngsters will need before they become independent and how much reproductive future the parent has left. Under most circumstances, the decision is not thought to involve how much has already been invested in the current brood—an argument known as the "Concorde Fallacy," after the ill-advised persistence in developing the supersonic transport plane long after it had been judged to be a money-losing venture[24]—but to focus solely on the parent's future options from the moment the crisis is first detected.

It follows that a jilted parent faces the same tough decision as one whose partner is killed by a predator or a virus. A voluntarily abandoned parent is in what Robert Trivers called the "cruel bind": it must either desert the youngsters as its partner did, condemning them to certain death and thus forfeiting the potential fitness they represent, or carry the full parental load by itself.[25] The vanishing mate may win a double victory, in terms of reproductive success, if it finds a new partner and starts a new family while its ex opts to finish raising the original brood.

The most common form of desertion involves the male partner taking off and leaving his mate in the lurch. Males, by definition, are the sex that produces the tiny but mobile sex cells (spermatozoa), so it appears at first glimpse that they have a very low initial investment in the production of a fertilized zygote. This is a potentially misleading view because to get the fertilization job done the male does not discharge a single sperm but tens of millions of them, and males can run out of sperm.[26] Looking at the situation more in terms of each sex's future opportunities, the higher incidence of male desertion may be tied more logically to that sex having broader alternative options than females.

Production costs aside, the asymmetry in size of male and female gametes was probably important in setting up a cascade of emergent sex-specific properties as soon as animals ceased doing all reproduction in water. The move to dry land and the concomitant evolution of the ability to fertilize eggs internally, in the portable sea of a female reproductive tract, burdened the female with possession of the zygote while leaving the male relatively free. In external fertilization, by contrast, the female typically sheds her ova first. A male fish, for example, often prepares an indentation in the substrate then courts a female until she lays in his nest. Immediately thereafter, he passes over the eggs, releasing sperm, but she has a brief opportunity during his ejaculation to swim off and leave him in possession of the family, effectively pregnant. Accordingly, single-parent care by males is far more widespread in fishes than in any other vertebrate group and has even taken some remarkable secondary steps, as described earlier for the pouch-breeding pipefish. Among invertebrates, the female giant water bug lays her eggs on the male's back, where they stick to an egg pad. The male then continuously carries the clutch, a clumsy and potentially risk-increasing burden that may weigh twice as much as he does. The male giant water bug, though, seems to be no fool. While the female is placing her eggs on his back, he interrupts after every second egg or so and re-mates with her, virtually guaranteeing that he is the genetic sire of all the eggs he lugs around.[27]

Apple snails are large freshwater neotropical mollusks that spend much of their time near the surface of shallow sloughs in southern Florida. They are favored as food by the few birds possessing the bill morphology and skill necessary to extract the body from its protective coiled shell. One of these predators is the snail kite, a crow-sized hawk with an exceptionally long and curved beak. It dines almost exclusively on the snails, tying its fortunes (and its behavioral options) to the availability of the prey population. These kites nest in vegetation a few feet above the water surface and can breed nearly year-round. Both sexes incubate the eggs and deliver freshly extracted snails to their nestlings. In about three-quarters of the nesting cycles, and only when water levels and snail populations are high, one parent deserts the family midway through the chicks' two-month dependency period, leaving the other to finish raising the kids. What is peculiar about this species is that the deserter can be of either sex, with mothers abandoning roughly twice as often as fathers.[28] This "ambisexual" desertion pattern means that Trivers's cruel bind can cut both ways. The remaining parent increases its delivery rate so that the brood of one to three chicks does not get less food, and fledging success is not affected. That is, these kites practice "full compensation" when abandoned.

So what determines which parent takes off? After abandoning its first family, the emancipated parent apparently finds a new mate and starts a second family. I say apparently because this has not been easy to document. However, one female that was tracked (she was carrying a radio transmitter) definitely found a new mate and was two weeks into incubating her second set of eggs when her ex-partner finished supporting

A snail kite delivers an apple snail to its brood of three chicks. In this species both parents typically care for the young, but either parent may desert the family if a second mating partner becomes available, leaving the other to raise the chicks alone. (Photo: Noel R. Snyder.)

the first family. By observing nesting pairs for more than two thousand daylight hours, Steve Beissinger detected a surprising trend: the parent that eventually deserted was almost always the one that delivered *less* food during the first few weeks after hatching.[29] This means the departure is unlikely to be a surprise for the partner about to be jilted. And that raises the issue of why such a parent does not see what's coming and turn the tables, sneaking off before its partner can do so. One critical component of mate-desertion in this kite is the availability of unattached members of the opposite sex: if a prospecting single bird appears in the neighborhood, *its* gender may determine which member of an earlier-nesting pair will cut down its feeding rate and eventually decamp from its current family.[30]

Abandoning one mating partner in order to take up with a new one is a familiar pattern to us humans. In the centuries since Henry VIII ruled England, many a public figure has opted—rightly or wrongly—to make such an exchange in the pursuit of happiness (or ego-fodder). In terms of Darwinian fitness, a new mate may actually offer tangible benefits of youth (greater residual reproductive value, more energy for parenting) or may simply be available (a hot, unattached snail kite of the opposite sex). If you extend that logic from mating to childcare, it raises the less familiar possibility of parents' producing more offspring than they expect to raise so as to have an array of youngsters to choose from. Do real families ever work that way? We're about to explore this question.

14

Upgrading the Kids

The average person thinks he isn't.

—Father Larry Lorenzoni

Among the traumas experienced by children at school (and no doubt revisited years later on psychotherapist couches) is the choosing of teams at recess. Two ultra-popular or talented players are anointed as captains and everyone else stands off to one side, waiting with pride or dread for their social status to be exposed publicly. Selections are made quickly at first, then with some deliberation. With just a few unchosen sheep remaining, their faces mask over, feigning indifference while awaiting the death sentence. It turns out not to be a fatal passage at all, of course. Certainly it is nowhere near the extreme faced by millions of animal and plant youngsters every year. The profligate overproduction practiced by the oak tree in your yard, cranking out bushels of acorns every autumn—as if striving to overwhelm the appetites of all jays and squirrels in the neighborhood such that a few seeds may actually graduate into seedlings—provides a local reminder that most offspring face very long odds from the outset.

We turn here to the question of whether there is any element of parental *choice* in such biological systems. Specifically, do breeding adults ever create extra offspring, above and beyond the core set they will support, simply to have an array of candidates from which they can select the most worthy? If so, how do parents assess worthiness? Must offspring face a test of some sort? Is there some mechanism of choice, requiring no intellectual processes whatever, such that an oak tree might be reasonably regarded as a choosy parent?

The basic thought here is similar in key respects to the insurance argument for parental overproduction, with the crucial difference being that insurance protects parents from a *numerical* error (raising fewer offspring in a breeding cycle than could be afforded) while this one focuses on offspring *quality.* If parents set their initial family size at an unrealistically high figure with the contingency of purging the weakest offspring later, they automatically raise the average vigor of the ones that survive.

If this sounds a bit like the way one chooses a mate, it should. One of the main processes of Darwin's theory of sexual selection, commonly labeled "inter-sexual selection" to distinguish it from same-sex competitions, concerns the way a prospecting adult can enhance its own fitness by being picky during mating decisions. By passing up partners that have less to offer (either of genes or wealth) and holding out for partners of high quality, an adult stands to produce offspring that will be more successful in surviving and, thus, in producing grandkids.

There is a vast scientific literature on mate choice, with considerable attention to what is really gained from heightened selectivity. To give but a tiny sample, in some species the rewards are tangible payments of resources: access to a particularly rich territory, perhaps, or even the presentation of a discrete food object in exchange for sexual favors. A gangly male black-tipped hangingfly must capture an edible insect and hand it over as a nuptial gift (payment up front) before the female allows him to mate: the bigger his bribe, the longer she is occupied eating the prey, the longer he mates (so the more sperm he transfers).[1] In the insect family known as balloon flies, males of some species similarly provide a fresh kill, while males in related species wrap the prey carefully in silk, apparently complicating the female's task and slowing her consumption, thus extending his copulation. In the most specialized balloon flies, males skip the whole business of catching prey and just make a ball of silk (the "balloon") for the intended mate. Other insect males manufacture nutritious body parts (such as wing pads) or special by-products (spermatophores) for females to eat.[2] The male black-widow spider sacrifices his whole body for the cause (*after* mating): he even performs a ritualized somersault that leaves him sprawled before her.[3] And the male preying mantis is said to increase his mating ardor if the female turns around and decapitates him during sex: beheading removes an inhibitory ganglion just below his brain, so the female gains both an early meal and a less inhibited partner.

In many other animals, no conspicuous goods are exchanged, but the behavioral components of female choosiness remain (some mates are rejected, others accepted), so it seems there must be other kinds of incentives. Certain individuals may be preferred as mates because they are perceived as being healthy (thus unlikely to infect the partner or their offspring), because they are likely to be carrying particularly valuable genes (and/or genes likely to be compatible with those of the choosing mate), or even because they are so good-looking that they are a good bet to produce handsome (and thus sexually successful) kids. Such arguments have been invoked to explain the evolution of aesthetic traits like the plumage extravagances of male peafowl,[4] birds of paradise, and pheasants.

The whole idea of being choosy in love can be restated in terms of how it affects offspring. After all, if a female takes in extra nutrients by ingesting a male's head, she uses those nutrients toward building their young. And if she rejects sexual partners because of perceived genetic shortcomings, it is on behalf of her offspring that she does so. Therefore, the idea of having another episode of choice later on, weeding bad offspring out of the brood, is not far-fetched.

This argument was originally developed for plants by J. T. Buchholz, who actually proposed "developmental selection" as a major evolutionary process, deserving of equal footing with Darwin's famous duo—natural selection and sexual selection.[5] It never caught on to that degree, but the basic potential for parental choosiness continues to show up here and there, especially in the botanical literature,[6] but occasionally with respect to animals as well. It travels under several aliases, but I usually call it progeny choice (to emphasize the parallels with mate choice). In fact, some biologists regard it as a form of plant mate choice because offspring fertilized by less desirable fathers (pollen donors) are discriminated against (selectively aborted) and the maternal resources thus saved are then available to half-sibs sired by preferred fathers.[7] I find this a reasonable position, requiring only that one broaden the context of mate choice to include post-mating decisions. Think of is as a "morning after" solution.

My favorite demonstration of progeny choice in plants features a species that has no English name so far as I know, thus must be addressed in all its Latin glory, *Mimulus guttatus.* Obviously, plants in general cannot control where their seeds or spores will end up after dispersal by extrinsic forces (wind, animals, and so on). This means they often land on soil with peculiar chemistry. If a population establishes itself in such a habitat, natural selection may eventually produce gene frequency shifts that enhance the ability to deal with the odd conditions. A population of *Mimulus* that had been thriving for a long time in soil contaminated with copper was used to test whether the mother (the maternal sporophyte) treated offspring differently depending on their paternity. The researchers dusted half of each female's ovules with pollen from donors living in similarly tainted soil (presumably bearing genes appropriately tuned by natural selection) and the other half with pollen from donors living in uncontaminated soil. That is, the mothers started out carrying both types of offspring. And they selectively aborted the offspring with "normal" type fathers and retained those from copper-tolerant ones.[8]

Progeny choice is also demonstrated by some plants that have two very different kinds of flowers, hence two very different sex lives. These species are hermaphroditic ("dioecious" in botanical parlance), meaning that each individual produces both male and female gametes. In such an arrangement, there are two general sources for pollen: a plant can use some of its own pollen to fertilize ova (self-fertilization) or it can obtain pollen from someone else (out-crossing). This pair of options lies behind the two kinds of flowers. The flowers used in "selfing" are small and insignificant, sometimes actually residing below the soil surface, which makes sense because they have no social commerce to conduct. You and I would not think of these as flowers at all, since they do not present the visual and olfactory feast normally associated with that word. Why should they, when they don't need to attract birds or bees? But they do have ovules and that is sufficient for flower status. By contrast, the big, showy flowers used in out-crossing are high on the plant and equipped with the usual accessories: large petals that serve as funnels, advertisements, and bribes for animal pollinators.

It follows that two classes of offspring can be produced, inbred and outbred, and herein lies the choice. If the parent has succeeded in out-crossing, it rejects the "selfed" offspring and invests in their more genetically diverse siblings. In short, the inbred zygotes are intrinsically inferior backups, serving only as a last resort; if not needed, they are discriminated against and jettisoned by their own mothers. (Let me pause to note that these stories of abscission all look like parent-offspring conflict, but they are presumably of the trivial "parent-wins" variety, since there is no reason to imagine that the maternal plant's interests are being abused in any way.)

In a great many flowering plants, selective abortion allows the parent to get rid of offspring (seeds or whole fruits) that have been damaged or that just happen to live on a branch that is having a bad season. For example, catalpa trees are chronically plagued by a highly specialized sphinx moth that lays large masses of moth eggs on the leaves. When the caterpillars hatch, they consume huge amounts of the photosynthetic surface that supports the tree. Having no way of anticipating which leaf areas are going to be defoliated (robbed of income), the tree has adopted the physiological trick of holding its tiny fruits in a lengthy stalling period until midsummer, when the moths pupate and fly away. At that point, the tree quickly aborts about 80 percent of its fruit crop (from the most damaged branches) and starts investing in the remaining few, which increase in weight ninefold in time for September seed dispersal.[9]

The parental practice of progeny choice is generally much better suited to plants than to most animals for several straightforward reasons. In the first place, lacking behavior to solve various problems, plants are much more prolific at overproducing. Consider a peach tree with its clouds of delicate blossoms in the spring and its far smaller crop of ripened fruits in the fall. Also, plants can finesse certain matters of timing better than most animals. There are, to be sure, animals that have the option of holding tiny offspring in lengthy periods of suspended animation called "diapause" (this is common in desert-dwelling species), but many plants have the additional habit of deferring most of their expensive investing until very late in the cycle. They pick their moment. One reason bananas taste so much better in the tropics than in temperate areas is because the ones marketed up here were picked green, shipped green, and typically sold half-green: the last shot of sugars normally goes into such fruit at the very end, which is also when the bright yellow color comes over the skins. This, of course, is also why homegrown tomatoes taste so much better than the objects sold under that name at the supermarket.

Having conceded that plants are more generally suited to the selective retention of certain offspring than animals, it is only fair to consider whether animals use progeny choice tricks, too. There are certainly cases of animal parents abandoning offspring and thereby condemning them to die. Wood storks in peninsular Florida desert whole broods of even advanced chicks if persistent rains occur during the usual dry season. This makes sense when you realize that these large wading birds support their families with a remarkable fish-capturing technique known as "tactolocation": they hunt by touch. Whereas herons and egrets hunt for fish in the same general habitats by visual methods, peering into the pools does not work well in highly turbid areas where the storks reign. These storks also use their toes to rake the muddy substrate at the bottom of shallow pools, increasing the murkiness and possibly forcing hiding fish to move around. Meanwhile, the stork pushes its slightly opened bill back and forth in the water, from side to side in a zigzag motion, as it wades about. Any fish that swims blindly into the edge of that bill witnesses one of the fastest reflex patterns ever measured for a vertebrate as the bill snaps shut around the prey. Then the bird lifts its head and tosses the fish down its gullet. Amazingly, blindfolded storks catch fish as well as controls do.

Wood storks breed during the dry season for an important ecological reason. As the weeks slide by without rain, water levels of standing pools drop, reducing the total volume and automatically concentrating all aquatic creatures therein. As a refined form of groping, tactolocation works best when prey are densely packed, so dropping water means that the extra costs of breeding become affordable.[10] Accordingly, if a lot of unexpected rain falls during a reproductive episode, completing the task may become a luxury that parents cannot afford, especially considering that they may live to breed twenty times or so. Sensitivity to out-of-season rain (actually to water depths and prey concentrations) thus appears to have evolved in support of prudence. When conditions become impossible, current offspring are sacrificed on behalf of future siblings.

The option of walking away from a family that is no longer cost-effective or that is utterly doomed is one of the advantages of raising young in a nursery that is not physically attached to (or within) the parent. These dramas unfold all the time. I remember being told as a child that I must never touch baby birds when I chanced to discover a nest because the mother would smell my scent and abandon them. This is hogwash. Over the past quarter of a century I have handled many hundreds of baby birds each field season and would have noticed if the Grim Reaper followed in my wake. Besides, most birds are believed to have weak powers of smell. But like most such parental admonitions, this one probably is good behavioral advice that might be reworded for candor: a nest found by a child, if not declared off-limits, is likely to be revisited many times (with many friends), and cats have an excellent sense of smell. Parent birds whose nest is likely to be found soon by a feline marauder may well cut their losses and desert the brood.

Sometimes it is possible for stressed parents to divide a family and abandon only some of the offspring when conditions are bad. In marsupial mice, more young are born than the six pouch teats can feed. To give birth the mother stands on all four feet with her hips raised, and the tiny, half-developed young race down to the pouch to claim spots. The ones that arrive too late are simply discarded on the ground.[11] Marsupials as a group are a bit like plants in this respect, featuring lengthy diapauses during which partly developed young delay further growth. In kangaroos and wallabies, for example, the nursing of young does not block a female from conceiving. If a mother with one joey in her pouch becomes pregnant, the younger sibling arrests development at roughly the 100-cell stage until the older one has been weaned. In grizzly bears, mothers with two cubs have been reported to tree one and lead the other away from its forsaken sibling.[12]

A remarkable study of Seychelles warblers by Jan Komdeur and his associates may also turn out to involve a progeny choice mechanism. Their central discovery was that parents seem to have control over the sex of the offspring they raise. This species lives in an unusual ecological situation. The Seychelles Islands are just south of the equator in the Indian Ocean, roughly a thousand miles out from Kenya. They are forested, tropical, and very isolated. This is the only warbler species there, and it presumably filled all territory space on the one island it inhabits a very long time ago. It has also evolved the cooperative form of social behavior known as a "helper" system, meaning that offspring may not (often cannot) disperse, but remain at home to help parents raise the next brood. The crowding is exacerbated by an unusually high rate of adult survivorship (over 80 percent). In short, this is an extraordinarily saturated habitat and, not surprisingly, family size is as small as it can possibly be: only a single egg is laid per breeding cycle.

Most of the helpers are daughters and there is a very finite limit on how many of these can be accommodated, a limit that varies with the quality of the family farm. On food-rich territories, parental success is enhanced by the presence of one or two daughters as assistants, but lowered if three or more remain at home; on poor territories, no helpers are affordable. Thus there seems to be some advantage to producing sons on poor territories and daughters on good territories that still have at least one helper vacancy. This is exactly what the birds do, raising 77 percent male chicks in poor areas, 88 percent female chicks where a helper would be useful.

The effect has even been demonstrated experimentally, through a conservation program that relocated pairs to two previously unoccupied smaller islands nearby. When pairs that had been living on poor territories back home (and making 90 percent sons) were moved to good territories on the new island, they switched immediately to the production of 85 percent daughters. Another sample of parents that had been producing sons on good territories fully loaded with helpers had their daughters removed (relocated to the new islands) and switched immediately to creating daughters.[13]

The question then arises, how can an animal (like humans) whose offspring sex is determined by the chromosomal makeup of the fertilized egg call up sons versus daughters, as needed? And the answer is . . . drumroll . . . we do not know. Perhaps the most realistic possibility requires only that mothers be able to *recognize* whether a given egg contains a male or female embryo, then discard any that are of the unwanted sex. If so, that trial-and-error process would constitute progeny choice.

There are, in fact, a handful of brood-reducing birds for which offspring sex does not seem to be random. In cattle egrets[14] and laughing kookaburras[15] males tend to emerge from A-eggs; females from B-eggs. In species whose nestling sons and daughters grow at substantially different rates, various explanations have been given for why one sex ought to precede the other and some field evidence has been marshaled showing that such patterns may exist.[16] Still, with birds and other vertebrates that use chromosomal sex determination, we continue to puzzle over how any special arrangements are actually handled by the parents.

The problem is much simpler for parasitoid wasps, with their haplodiploid system of making male and female offspring. Not only is their mechanism well understood, but it is also much tidier, since the choice is made well in advance. All that one needs to make sense of it all is a motive, a fitness incentive, for deviating from a fifty-fifty sex ratio. Parasitoids base offspring sex on the size of the host, just as they do family size. For example, a female *Lariophagus* wasp that has paralyzed a granary weevil host introduces an excess of (diploid) daughter eggs if the weevil is large and sons (haploid) if it is small.[17]

A major impetus for Darwin's theory of sexual selection was provided by the exaggerated ornamentation found on many male birds, including bright colors, elongated tails, and sometimes even extra-body confections like the exquisitely decorated display edifices erected by male bowerbirds. These visual excesses (plus rococo acoustic parallels found in avian song) were postulated as having evolved because prospective mates found them attractive, though most explanations also required that there be a link between the stimulatory trait and some truly important, if less conspicuous, aspect of male quality. To give just one example, it has been suggested that the red wattles and comb of the male red jungle fowl (the ancestor from which our chickens were domesticated) indicate the possessor's ability to resist blood parasites.[18] A glowing red ornament may give highly reliable information because the presence of plenty of red blood cells cannot be faked. And experimental studies have indeed shown that nematode-injected male jungle fowl have paler ornaments and are eschewed by hens.[19]

As before, we ask whether something like that might not be involved in the way parents regard their offspring. If adult birds are sensitive to flashy cues when making mating decisions from an array of suitors, it makes sense that they might attend to comparable sources of information from their own young a few weeks later. Becky Kilner has shown that canary chicks may reveal their level of need to parents via the pink coloration of their mouth linings, which "flashes" rapidly at the beginning of a begging bout. The hue and brightness achieved vary with a chick's hunger level: when the chick is temporarily deprived of food, the color deepens. When parents were offered a choice of two nestlings, one of which had its mouth artificially reddened, they showed a clear preference for feeding the brighter mouth.[20]

But wait! If that sort of thing is going on in many species, should we not expect to find some cases where the evolutionary arms race among competitors has produced some traits a lot more ostentatious than deep pink mouth linings? Is there nothing comparable to the peacock's tail in youngsters' signals to their parents? In one sense, we might not expect to find much extremism in offspring signals. Whereas male peafowl are competing against unrelated sexual rivals and have their own methods of avoiding predators, nestling songbirds are pitted against their closest genetic relatives and face a shared disaster if showy signals catch the attention of a bobcat. There are some interesting possibilities under consideration here, including the one about offspring begging being a form of blackmail to coerce parental investment. And certainly when we relax the stipulation that nestmates be close kin, fancy situations like the cuckoo begging at the rate of eight warbler chicks can arise.

The most radical-looking hatchling birds that I know of are young American coots. Whereas the downy plumages of most avian nestlings are somber mixes of brown and tan, camouflaged to escape visual detection by enemies, the marsh-dwelling young coot sports an orange anterior, a bright red bill, and a bald red pate. These gussied-up youngsters are mobile, traveling around the parents' jointly defended territory for their first month, and they compete with their siblings for food collected by their parents. After one week of being tended en masse, the brood is divided between mother and father. The coot chicks' begging displays are not vocal, but feature postures that exhibit the bright plumage and bald crown. And at the age of three weeks the chicks begin to shed their ornamental feathers, so those who survive to independence at the age of one month are dark and plain. All these lines of evidence suggest that parental preferences for punk coloration are ephemeral and must play their role early on.

Bruce Lyon studied coot families in the beautiful wetlands of central British Columbia, exploring whether parental favoritism might be responsible for the hatchlings' garish appearance. He found that parents seem to exercise complete control over which chick receives each food item and that they make a point of getting more food to the youngest member of each half-brood.[21] This is a brood-reducing species (between one-third and one-half of all broods lose some chicks to starvation) without sib-fights, so parental behavior truly shapes which siblings do well and which suffer.

To explore the possible role of bright plumage, half the chicks in each of twenty-one broods were experimentally stripped of their bright coloring (only the tips of the feathers are orange, so clipping them at mid-feather leaves a black down that insulates the chick), while their siblings remained bright orange and served as within-treatment controls. There were also two between-group controls: twelve broods had all the chicks clipped to look dark, while another fifteen broods remained unclipped and orange. Those two groups were treated very much the same by parents, so the focus of interest is on the mixed treatment of clipped versus unclipped chicks in broods where both were present. In these broods, unclipped orange chicks were fed more, grew more rapidly, and survived more often than their less colorful siblings. The effects were strongest for the youngest chicks in each group. If they retained their orange plumage, hatching order did not affect survival at all: chicks that were five days younger than their broods' eldest survived at the same high rate (80 percent) as A-chicks; but if they had lost their orange feather tips, survival was only half as likely. In short, the flashy plumage of baby coots does seem important to their parents and may have evolved in response to parental aesthetic biases of some sort. This seems to be quite a

An American coot seizes one of its offspring by the head in a punishment known as "tousling." In this and related species, such punishment seems to be largely directed at the larger siblings in a brood and to enable parents to feed smaller chicks preferentially. In the European coot, tousling can prove fatal. (Photo: Bruce E. Lyon.)

clear behavioral case of parents actively making choices among their offspring.

The functional reason for such biases—how they might enhance the parents' fitness—is less clear. One possibility is that a chick's condition is revealed by its bald crown (where circulating blood can be viewed, as with canary mouth linings and rooster combs), but that does not explain why parents prefer orange feathers. Perhaps parents ordinarily work hard to counteract the effects of hatching asynchrony (which probably cannot be avoided in a chilly climate where the first-laid eggs need to be protected from the cold). That, in turn, may entail selective boosting of the youngest chicks; but there may be cues that parents use to drop certain chicks from the high-priority list. The favoritism of parents in the closely related European coot has been reported to include fatal thrashing ("tousling") of certain young.[22] Interestingly, that activity usually kills only the senior siblings. American coot parents also tousle their chicks, but seemingly are less violent in that the abuse has not yet been observed to be fatal.

For any case of parental overproduction to count as progeny choice, the key questions are whether parents can improve their own fitness by screening their offspring on the basis of quality and whether the gains thus obtained are likely, on average, to exceed the inevitable costs. In other words, is such screening affordable?[23] As discussed earlier, many plants are beautifully adapted for reaping such benefits because their offspring (seeds and fruits) can be retained for a long time at a very low unit cost and the surcharge a parent must pay to dismiss rejects is also very low: the maternal sporophyte simply lets go of the unwanted offspring and gravity removes the problem. In addition, because plants lack behavior, including the power to choose mates, most receive male gametes from a much wider range of donors, potentially representing a much wider spectrum of genetic quality, than is the case for animals. So the interaction between the male and female genes in the resulting zygote is similarly likely to generate greater variation in offspring phenotypes. When your love life is being conducted for you by the winds (or bees) of fate, a veto option is especially valuable. And as long as the maternal plant retains and supports her offspring, she is in excellent position to evaluate their progress and worthiness.

Eagles, egrets, and other siblicidal birds have also been suggested as candidates for a version of progeny choice sometimes called the superchick hypothesis.[24] The notion here is that parents create an extra chick (rarely two extras) and let the kids fight it out, using sibling rivalry to identify especially vigorous offspring that are, therefore, especially deserving of the lengthy and expensive investment to follow. The victor, it is sometimes argued, also gets "practice" for the rough-and-tumble life it will lead in adulthood, when it will have to kill unwilling prey in order to eat and to defeat dangerous adult conspecifics in order to obtain and defend a breeding territory. One obvious problem for applying progeny choice logic to these species comes from the parents' asymmetrical manipulations of sibling rivalry—if one chick is always younger and smaller than the others, its innate vigor may not be detectable in sibling battles—but we might envision an occasional marginalized chick so gifted, so fiercely deserving, that it pulls off a *coup d'état* against long odds and reverses the dominance order. Hence the term "super-chick."

That super-chick scenario starts unraveling when scrutinized closely (and preferably mathematically).[25] The problem is that if the parent's goal is to identify which of its offspring is a genetically exceptional specimen, a feathered Leonardo da Vinci, the worst tactic is to cloud the comparisons with nongenetic complications.[26] If a track coach wants to determine the best miler on his team, it will not help to give some runners suitable footwear and make others don hip-waders. Similarly, if an egret parent needs to evaluate which of its three chicks is intrinsically the weakest (and therefore most expendable), it gains nothing from making one hatch several days after the other two or shortchanging its yolk steroids. Buchholz knew all this back in 1922; he explicitly noted that for an effective screening mechanism all siblings must get an even start and compete under identical conditions.

For siblicidal birds, thus, parents' incentives for overproduction are unlikely to involve progeny choice. Imagine a privileged A-chick with some "bad genes," say a few alleles that render it more vulnerable to pathogens, so that it cannot prevent a healthier junior sib from usurping its place in the core brood. In a series of such families, the parental handicaps imposed on junior sibs via asynchronous hatching would eliminate many that happened to be only somewhat healthier. In all those cases the parents would then have to pay a variety of associated costs (having to raise the defective A-chick most of the time, having that survivor further weakened by the effort of killing a challenging nestmate, and so on).[27] For each junior super-chick that succeeded in overcoming parentally imposed obstacles, many "almost-super" others would fall by the wayside.

To date, then, we understand rather little about the role progeny choice plays in the evolution of family size. Our ignorance may be due to its simply having been overlooked for the eight decades since

Buchholz proposed its importance, or to its not really being an important evolutionary process. The increasing attention to various forms of parental favoritism that are behavioral (offering food to certain offspring and not others, selective abandonment and infanticide, and so on) may soon lead to deeper contemplation of these possibilities.

15

Together Again

I have found the best way to give advice to your children is to find out what they want and then advise them to do it.

—Harry S Truman

By now it should be clear that relationships within families are a diverse, complex, and dynamic potpourri of both positive and negative elements. Hamilton's rule revolutionized our approach to both types. First, it helped us see the extended family as an important key to understanding altruism (and those explorations led to a fuller appreciation of forces other than kinship that can also promote cooperation). Second, it provided a theoretical framework for studying puzzling cases of extreme selfishness, where even close kin may be sacrificed. Along the way, many topics in behavioral and evolutionary ecology enriched, and were enriched by, the work inspired by inclusive fitness theory. In this book the focus has been on family strife (sibling rivalry, sexual conflict, and parent-offspring conflict), with special reference to the unpredictable nature of key resources such as food. A second focus has been on the close connections between life-history traits such as family size and behavioral adaptations. All such traits are subject to the unsentimental shaping powers of natural selection.

Inasmuch as nearly all the research described in these chapters has been conducted in the past twenty-five years, progress has clearly been diverse, rapid, and exciting. Still ahead lies a full understanding of shared and conflicting interests between family members. We must learn more about this fascinating social microcosm. And so I will focus this last chapter on a few unusual systems in which shifting roles played by various family members strike me as especially complicated.

* * *

Let us start with pigs. You may conjure up images of pink, sweet little piggies or ponderous and malodorous adult porkers: both visions are apt. The life of domestic swine begins inside a sow, of course, and the unfairness of it all gets going right away with the implantation of embryos in the uterine lining. As in a pronghorn womb, some parking places are better than others. Piglets at birth show variation in size and disposition that result from having developed at the more crowded ends of the uterus, and perhaps from having developed between two members of the opposite sex[1] or from patchiness in the uterine blood supply.[2]

Pig litters are also exceptionally large for an ungulate. The world into which they emerge at birth contains another salient unfairness: teats toward the anterior end of the mother's body provide more milk than those at the rear. Not surprisingly, the biggest pigs can be found riding in first class; the runts in coach (hence the colloquialism, "sucking hind teat"). Thereafter, piglets tend to return to their favorite teat (perhaps guided by the odor of their own saliva, as mouse pups are).[3] As with asynchronously hatching birds and dozens of other sibling rivalry stories related in earlier chapters, this inequity is self-sustaining: the rich get richer and so on. This is a different type of situation, though, and not merely because we are dealing with a mammal. Piglets belong to the small fraternity of infants that carry weapons.

Piglets are born with unusually precocial dentition, in particular, fully erupted eyeteeth that jut out from the lower jaw at a low and sideways angle. Farmers routinely clip these out of concern for the sow's udder, which is to say that these teeth have long been regarded as an odd and inconvenient feature and no one has paid much attention to why they exist in the first place. David Fraser, a research biologist with Agriculture Canada, noticed that piglets use their eyeteeth in defense of preferred teats.[4] He performed an elegant experiment in which he clipped them only from heavier littermates, leaving the smaller siblings with their weapons intact. This leveled the playing field considerably and enabled the lighter piglets to grow as rapidly as their larger sibs.[5]

The evolutionary significance of all this is a bit hard to judge, considering that the research swine herd at Agriculture Canada is domesticated (that is, from ancestral stock pruned by human preferences for many centuries in the process Darwin called artificial selection) and lives in clean research barns. It may be that piglet face slashing under more septic wild conditions leads to serious infections. Alternatively, it may be that this weapon and its use all evolved within the context of domestication, as a consequence of farmers wanting larger and larger litters. Obviously, somebody needs to go have a look at wild pigs, like warthogs and peccaries.

Fraser has pointed out that, at least in current farm practices (where sows give birth in small farrowing crates), the real danger for a runt is that it must gamble in order to grow. Because its hind-teat gives less milk, it must spend more time trying to nurse, which jeopardizes its survival simply because of its mother's enormousness. Any piglet caught beneath a sow that happens to roll over is doomed. Fraser invites us to imagine ourselves naked and sharing a small unheated cell with a lactating female elephant that is simultaneously our only source of food and warmth.[6] Get the picture? For pigs, then, it is the great bulk of the mother that most directly threatens the lives of the smallest babies, but the special weaponry of the siblings' teeth forces them into some extra risk-taking. The mother's other role, of course, lies in the unevenness of her milk outlets.

There are other species, too, in which very young offspring possess special weapons that may exist solely for use against their nurserymates. We have already met the parasitoid wasp *Copidosoma,* in which some of the female clones develop gigantic jaws that they use to cannibalize their brothers and adjust the brood sex ratio. Indeed, those precocious larvae themselves might be thought of as a weapon belonging to their reproductive clones. And we know about the young cuckoo's behavioral traits that help it evict nestmates, though the victims in that case are not siblings of the killer. Among the other remarkable brood parasites are the honeyguides (named for their habit of directing honey badgers and humans to the hives of honeybees), a family of birds that lay eggs singly in the nests of woodpecker and barbet hosts. Newly hatched honeyguides use nastily hooked bills to macerate the resident chicks and eliminate the competition. A Southeast Asian bird, the blue-fronted bee-eater, does a similar number on its true siblings. It has a hook in its bill at hatching that straightens out later on, after the siblicide is done.[7]

Spotted hyenas are a bit like pigs in having impressive teeth (fully erupted incisors and canines) at birth, which they use to dominate and often kill their sibling den-mate.[8] Adult female hyenas are larger and more aggressive than the males, which they dominate in the pack's social structure. The female carries extraordinarily high levels of circulating androgens and develops an oversized clitoris, the "pseudopenis," through which the birth canal passes. The suggestion has been made that much of hyena sib-fighting might be a side effect of selection for female aggressiveness and some early field evidence also suggested a sex bias in the siblicidal activities, as it appeared that same-sex pairs of cubs were more likely to produce a fatality.[9] Other researchers have found no such sex bias in hyena siblicide, but argue that it appears to be driven by food shortages (like the siblicide of many birds).[10] It is thus unclear whether the hyena's siblicidal specializations are related to the masculine nature of the females, but the matter is under vigorous study.

The laughing kookaburra of Australia is an aberrant kingfisher with a different sort of complex social life. Like Seychelles warblers, kookaburras live cooperatively with one monogamous breeding pair supported by older offspring that have not left the home territory. Unlike the parents in many such systems, kookaburra parents reduce their own investment in a given brood in direct proportion to that provided by the helpers (mostly males in this species),[11] such that the typical three-chick broods receive little or no food bonus from having extra hands on deck. Early incubation by the mother produces the familiar pattern of asynchronous hatching, as if setting up the brood for future pruning. And, sure enough, there is lots of early-life mortality to face, such that nearly half of all eggs do not produce a fledgling. About a sixth of the offspring fail as eggs (because of infertility, accident, predation, and so on) and then two distinct waves of sib deaths bite into the rest. And bite is the right word, because these nestlings, like the bee-eaters, have temporarily hooked bills that they use early on to torment younger siblings. So fixated are the bullies on grasping the heads of their victims and shaking them violently that the action may continue even when researchers lift them out of the nest and carry them to the ground for various measure-

A young laughing kookaburra bites its nestmate in a process of fighting and intimidation that allows the bully to take the most advantageous position in the burrow, where it can intercept food delivered by the parents. (Photo: Sarah Legge.)

ments.[12] Sibling aggression, detected mainly from the presence of flesh wounds, contributes to a wave of C-chick deaths in the first few days. A second wave of brood reduction occurs a week or two later, this one accompanied by fewer wounds (the bill-hooks are gone). These delayed deaths are mainly due to starvation, though sibling aggression may still play a (less easily detected) role, when family food requirements are at the zenith.

It is all well and good to speak of parents as facilitating sib killing, but how often do the mom and dad provide the actual muscle? After all, we have been aware for over half a century, since David Lack published his paper on hatching asynchrony in 1947, that some of the things parents do place certain offspring at a disadvantage they are unlikely to overcome. But if parents are not perfectly evenhanded, and if they are not even working at their own maximum levels to ensure the welfare of the whole brood, we must ask why they seem to leave the messy job of execution to siblings of the designated victim.

We have encountered some cases of flat-out execution-style infanticide in coots, with parents sometimes tousling individual offspring to death. Of course, the great majority of infanticide in nature offers no special mystery for the simple reason that the victims are not related to their executioners. For example, when a male Hanuman langur usurps a troop of females—the Banderlog of Kipling's *Jungle Book*—he apparently kills the infants already present.[13] In the ornithological literature, accounts of infanticide by conspecific adults are pretty sparse.[14] One of my all-time favorite titles from the journal *Science* (generally not a great source of frolicsome titles) is H. L. Ward's 1906 eye-catcher "Why Do Herring Gulls Kill Their Young?" I sprinted to the library when I first found this reference, but was disappointed to read that the author had no idea whether the perpetrators were killing their *own* chicks or somebody else's. His observations were made during a casual stroll through a dense gull colony, an activity that sends hundreds of youngsters scuttling recklessly through the grass and bushes to avoid the human. (The carnage that followed, which has been ascribed clearly to neighboring adults, may well have been the first published clue that gull parents have keen kin recognition abilities.) Now, nearly a century after Ward's paper, herring gulls are famous for cannibalizing their neighbors' chicks when given half a chance. In one dense colony it was even reported that individual adults that specialize on a diet of local nestlings sometimes had the tactic backfire on them when they failed to complete the killing process. Some of the stockpiled "carcasses" managed to recover from their initial thrashing and promptly moved into the cannibal's home brood, paradoxically adding to the hungry mouths to be fed.[15]

If we restrict ourselves to species in which parents are known to kill their own offspring, we find some cases that are fairly understandable (such as wood storks facing unseasonable rains that jeopardize their livelihood) and some puzzles (such as royal penguin mothers that kick out their small first eggs). Stephen Jay Gould, in one of his famous essays for *Natural History,* described what might have become a classic example of parental infanticide when he described the "guano ring" of blue-footed boobies.[16] This booby (which lives up to its name by having sky-blue feet that are held up in deliberate alternation during courtship displays) has much sibling aggression and brood reduction. When Gould wrote about it in the early 1980s, little attention had been given to siblicide for any bird, and he had recently visited a nesting colony in the Galápagos, where he had spent only a few minutes casing out the situation. His impression was that the execution of the junior sibling began with a modest bit of bullying by its nestmate that caused the victim to flee out of the nest, specifically out beyond the circle of feces that the fastidious parents had squirted away from the nest during their weeks of incubation and brooding, but that the parents prevented the victim from returning home and thus condemned it to death. Gould surmised that parent boobies do not regard anyone crossing that line lightly and so, once evicted by its sib, the victim is denied salvation by its own parents' rigid enforcement of the sanctity of the guano ring.

I recount this explanation with a measure of amusement because Gould, with all due respect for his many contributions to evolutionary biology, took behavioral ecologists to task on several occasions for invoking natural selection excessively. Recall from Chapter 6 his clever comparison of adaptationists to believers in Kipling's *Just-So Stories*. Were he still among us, I have no doubt that he would find a few just-so stories in this book. But with the boobies he crossed the line himself. On the basis of exceedingly little data, he wrote a delightful essay on siblicide that contains the mortal vulnerability of all scientific interpretation. That is, it is *testable*.

David Anderson, who has spent much of his professional life in the Galápagos Islands studying these boobies, suspected that Gould's story about the key role of parental intolerance was bogus. And so he performed a simple experiment. Tying a length of fishing line to the leg of a booby chick and retiring a distance away so as not to be a complicating factor (even to these ultra-tolerant birds), he waited a few minutes and then slowly pulled the chick out of its nest and beyond the guano ring. According to Gould's hypothesis, this should make it unwelcome when it tried to return. But every trial at each of several nests produced the same response: the parents always allowed the chicks back into their nests.[17] Detailed studies of what really happens to blue-footed booby victims, conducted both in the Galápagos and off the western coast of Mexico by Hugh Drummond and his students, indicate that parents are not active participants at all after hatching.[18] The smaller sib is evicted, often many times, by its elder sibling and often perishes because of abuse by neighboring adults or is snatched by a predator lurking about the colony.

Working with a student on Heermann's gull, which is visually peculiar (dark body with white head instead of the reverse), Drummond found that in this species the parent does contribute actively to the physical abuse of the youngest chick, pecking it and driving it out of the nest.[19] Though it does not die as a direct result of parental attacks, being expelled from the nest also means losing its access to shade (beneath the parent), which is fatal.

An even more direct role in offspring destruction is played by parents among burying beetles, an attractive group of large insects that practice a funereal trade. When a mouse or small bird dies from some cause other than predation, its corpse is often found quickly by adult burying beetles, which use it as a resource base for reproduction. A male and a female, presumably meeting for the first time over this prize, defend it from other beetles while preparing it for interment. They denude the mouse of fur and shape it into a ball, which they push into a small excavation and cover with soil. They then mate and share a defensive vigil over the developing eggs. When the larvae hatch, parents regurgitate mouse flesh to support their growth. The female calibrates family size to match the size, and more specifically the volume, of the food ball. If tiny balloons are inflated within the cadaver by a curious experimenter, falsely expanding its apparent size, the mother lays more eggs.[20] And if she errs in her estimate of how many larvae can be supported by a given ball, a familiar dilemma arises. In the case of burying beetles, parents respond to such supply-demand mismatches by the simple expediency of cannibalizing some of their young and feeding that regurgitate also to the others.[21]

Rodent mothers are also known to cannibalize their own litters, an action that is usually explained in terms of that litter being doomed anyway (say, about to be discovered by an unstoppable predator). Sometimes mothers consume sickly individuals and spare the rest, a form of progeny choice. Furthermore, male rodents, like langurs, practice infanticide when they encounter a nursing litter that they did not sire. Indeed, it has been shown that the new male can keep track of how long he has been mating with a given female and will cannibalize pups born too early for him to have been the sire.[22] Pregnant female mice that are exposed to a new male (or even soiled bedding bearing the odor of a new male) are likely to abort their litters and resume cycling.[23] Similar spontaneous abortions have been reported for wild mustangs after a herd of mares is won in combat by a new stallion.[24] Purging of unborn offspring due to the presence of a new and potentially threatening male (called the Bruce effect after its discoverer, Hilda Bruce)[25] may be more widespread than we know, since it is extremely difficult to detect under field conditions.

The evolutionary significance of langur-style infanticide has been much discussed in terms of Darwinian sexual selection (traits arising because they attract mates).[26] The newly arrived male has no genetic stake in his young victims, so there is little to hold him back, and the killing is likely to boost his fitness by accelerating his own sexual access to the mothers. That is, by removing the suckling infant, he eliminates the hormone-mediated block on female estrus, returning the ex-mother to a state of sexual receptivity more quickly. Such infanticide is presumably harmful to the mother's fitness (except, perhaps, in rare cases where the original male was unable to defend his family because of a genetic defect), but is imposed upon her anyway.

The Bruce effect, by contrast, rests on the sensitivity of females to the odors of a new male, and it is the female's physiology that responds to terminate the pregnancy. Whereas male infanticide is apparently imposed on her litter, this response requires a more female-centered explanation. What might a mother gain from dumping her litter under these circumstances? Is she simply making the best of a bad situation? To phrase this issue a bit more formally, and in a way that parallels the discussion of the queen bee's control of workers, how would a mutation fare if it produced female mice that simply ignored the chemical signal (the odor of the new male's urine) and did not abort? The mutation's real cost is delivered through the consequences at the bottom line. If a new male has established himself in her area and is likely to slaughter her litter after they are born, an early abortion saves her the time and energy she has not yet invested in old male's pups. In short, the price she pays for doomed pups can only go up if she delays.

I want to close these accounts of parental infanticide with a point about the way many such phenomena ever come to light. Much of what we now find comprehensible was not even imaginable just a few decades ago. Science proceeds in fits and starts, sometimes with a new technique that allows us to measure something previously out of our reach (such as relatedness), sometimes with fresh thinking and an argument that makes so much sense that it inspires us to look at things differently. And sometimes it is simply that the old explanations come to be at odds with the accumulated facts. A topic like parental infanticide is greeted at first with a wall of denial. Being both rare in most animal populations and viscerally abhorrent to human observers, it is likely to be underreported when first noticed. The initial reaction is to assume it is unrepresentative, a trivial pathology.

I became interested in the problem of underreporting when an ornithological journal editor contacted me several years ago to ask if I would take a look at a manuscript, with photographs, that described a single incident of parental infanticide in a large European bird, the black stork. The evidence was very straightforward and highly suggestive. Two Polish brothers, Grzegorz and Tomasz Kłosowski,[27] excellent nature photographers, were observing a brood of five nestlings from a blind. A visiting parent had just fed the brood on regurgitated fish and rested for twenty minutes, when it suddenly seized the smallest chick by the head and threw it over the rim. The chick landed on the ground ten meters below and died immediately.

Not a lot can be inferred from an anecdote like this. The attacking adult was not individually banded or identified by sex. The case for it being the victim's parent is thus circumstantial, based on the fact that black storks nest in considerable isolation deep in the forest (averaging more than two miles between nests). And the victim had just eaten food delivered by the killer. It seems highly improbable that a strange adult would fly a great distance and then feed the family before committing an anonymous infanticide. What I find so interesting about the incident

A black stork that has just fed its nestlings suddenly grabs one *(left)* and throws it over the nest rim *(right)* in a fatal eviction. Such execution of an adult's own offspring has been well documented in only a few species, but was considered unthinkable until recently. (Photo: G. and T. Klosowski.)

is that the Kłosowski brothers probably would never have reported it at all if the biology of infanticide had not become scientifically acceptable. As it was, they took the photographs on 18 June 1978 but did not publish them until encouraged to do so twenty-four years later.

The highly social acorn woodpecker of central California, which has been studied closely for over three decades by Walt Koenig and various associates at Berkeley, shows another case of offspring destruction by (known-to-be) close kin.[28] Generation after generation, birds in this population spend their whole lives with their extended families on large territories that contain a remarkable larder. Adept at woodworking, these birds chisel millions of small round holes in the trunks of large trees (and in barns, for that matter) and pack a single acorn in each hole. Year after year, acorn crops are harvested nut by nut and placed in dry storage for later meals. The holes are maintained and defended by the resident flock. Breeding is done in a communal nest cavity, and there is intense competition among the female kin to have their own eggs incubated. This takes the form of stealing about a quarter of the eggs from the nest, which are then destroyed and typically eaten. The *pièce de resistance* is that sisters commonly share the meal, sometimes with the egg's own mother.[29]

The basic structure of an acorn woodpecker flock—male group on one side and female kin group (not related to the males) on the other—is also found in African lions. The core of the pride is a team of up to twelve lionesses (sisters, cousins, aunts, nieces), to which two to four allied males attach themselves, arriving without invitation and, indeed, having to fight their way through the previous multi-male occupation force. The takeovers are spectacular events, often beginning with several days of roared challenges by the visiting team and bellicose responses from the home team, and eventually building to a crescendo if the intruders do not leave. A deadly battle ensues until one side is defeated. If their coup succeeds, the members of the new coalition move in and dispatch most of the nursing cubs quickly. Mothers may try to protect their cubs, but even a powerful 265-pound lioness is dwarfed by the 400-pound male.

A curious and subtle thing happens next. Though it is true that the ex-mothers resume cycling much sooner than if their cubs had been spared—so the male incentive for infanticide seems reasonable—females at such a time are notably less likely to become impregnated than at other times. They definitely resume cycling and soon become sexually receptive to the new males (so there is no question of their "punishing" the new coalition by withholding sexual favors), but those early matings simply do not take. At first this seems physiologically spiteful of the females, since they themselves are gaining no fitness during this time either. But another look at the situation reveals a benefit for the females that is surely *not* shared by the new males. Consider that the freshly victorious platoon of male warriors has just fought a vicious battle. They have been slapped, slashed, and bitten by comparably massive defensive gladiators fighting for their lives and their reproductive futures. The tooth enamel of felines, both large and small, is etched with tiny grooves harboring bacteria that make cat bite wounds notoriously septic. Low-grade infections suffered by the new males may contribute directly to their reduced potency. So one possibility is that the males are temporarily impotent. Alternatively, the reproductive physiology of the female may be imposing a brief time-out period. From the females' perspective, pride takeovers are very expensive events, as all the investment they have poured into existing cubs is summarily discarded by the victorious males. By stalling on the next pregnancy for a while, females determine whether the new coalition can recover from their battles and provide strong protection against the next group of marauding challengers, which might show up at any time.[30]

In 1926 Einstein wrote, "It is theory which determines what we can observe"—an assertion that must be true in a far more literal sense for subatomic particles than it is for behavioral ecology. Many of the actions that whole bodies perform can, after all, be described and then contemplated *a posteriori*. Classical natural history and ethology relied primarily on the observe-first-then-interpret approach. That remains an important source of material, but it is now complemented by sophisticated modeling that can steer us toward phenomena we might never have stumbled across by chance.

The study of siblicidal brood reduction sits astride both traditions. The earliest clues I know about can be found in Aristotle's *History of Animals* (book VI, chapter 6), where he described an eagle "that lays three, hatches two, and cares for one." He even provided a tiny hint at insurance logic ("though occasionally a brood of three has been observed") and assigned considerable blame to parental favoritism ("the mother becomes wearied with feeding them and extrudes one of the pair from the nest") before sailing completely over the boundaries of plausibility ("the phene [a vulture] is said to rear the young one that has been expelled"). The facts may be a bit muddled over who evicts the victim and what becomes of it, but he certainly was on the right track. Still, the subject remained at the level of natural history for at least two millennia, without notable progress in teasing apart its mysteries, until we built a theoretical framework that made sense out of it.

Elegant theory, though, must not become an end in itself. Devising a plausible explanation for something legitimizes further study by suggesting that the descriptive information already on hand (plus any clues available from tests of other ideas) may be part of something else, perhaps something more general and more coherent. As such, it sharpens the intellectual appetite and helps suggest how a methodical exploration might proceed. It provides key points that need critical testing. But until such tests are done, the hypothesis is only a hunch (and hunches are usually wrong).

Whether one starts with a gem from natural history or a lovely piece of math, curiosity about how things really work has to take over. What is perhaps underappreciated about theory as a jumping-off point is that one is never sure where that jump is going to land—and it often lands in several different places. As noted early in this book, the main theoretical foundation for studying siblicide was Hamilton's rule, the simple inequality ($br - c > 0$) that specifies the limits for altruism (and, when not satisfied, for selfishness) in terms of effects on the fitness of close kin. It has helped us understand much more about many cooperatively breeding vertebrates (such as Seychelles warblers, acorn woodpeckers, kookaburras) and insects (bees, wasps, ants) in which many mature adults seem to forgo personal breeding to assist their kin. Eventually, scrutiny of these apparently altruistic systems showed them to be riddled with selfish elements. Hamilton's rule embraces all of the above.

One of the first spin-offs from the recognition of kinship's central role in sociality came as a logical prediction. If altruistic acts can generate fitness gains when directed toward kin but not when directed toward non-kin, then one might expect natural selection to have shaped nervous systems capable of discriminating between these two classes of conspecifics. In the 1970s a demonstration of kin recognition was made by Bruce Waldman, who tested whether tadpoles behave differently toward fellow tadpoles from their own egg mass than toward those from other egg masses. This question would not have crossed anyone's mind before Hamilton. Nobody would have cared about something as esoteric as tadpole kin recognition. It became an important topic *because* a more general principle already existed.

Waldman explored whether American toad tadpoles preferred to associate with siblings more than would be expected by random chance. To do this, he captured adult pairs while they were mating, took them into the lab, and let them spawn in separate buckets so that he could hatch the resulting broods in isolation. Then he soaked some sibships in red dye (to stain the less pigmented edges of their fins) and others in blue. Taking a sample of red-finned siblings and a sample of blue siblings, he placed individuals randomly in a larger test pool and gave them time to sort themselves out, thus revealing that they did so according to relatedness. He carefully controlled for the staining procedure by dyeing half the members of some broods red and the other half blue, which showed that they were not using fin color as the basis for association. In short, he showed that tadpoles have some other means for identifying their siblings.[31]

This exciting news led immediately to a much tougher question: So what? It was not obvious how the tadpoles' ability to recognize kin could shed light on the bigger issue, the evolution of altruism. What might even the world's most generous tadpole possibly do for another tadpole? What form could the expected self-sacrifice take?

Various creative suggestions were made and pursued,[32] but the matter sat in limbo for several years before the central issue came to be restated more plausibly in terms of what an altruistic tadpole might *refrain from doing* to kin that it would happily do to non-kin. This besmirched rendition of the Golden Rule—Do not do unto certain others what you don't want them to do to you—had the merit of fitting neatly into a well-known piece of tadpole natural history. Some tadpoles are notoriously cannibalistic.

One assemblage of toad species seemed especially suited for studying whether kin recognition might be used adaptively during cannibalism. The spadefoot toads live across much of North America and have been studied mainly in arid parts of the Southwest, where they capitalize on sporadic episodes of heavy summer rain filling temporary puddles and gulches. When a half-inch or more has fallen and the low areas begin to fill, one or two male spadefoots begin to call and their noise attracts others until hundreds of males are producing a tumult. This attracts females, traveling alone but converging on the chorus. Upon reaching the water, a female swims about until she bumps into a male, causing him to seize her around the hips and hang on. (Males are quick to grab and frequently mistake each other for females, each trying to achieve the dorsal position on the other: when one finally succeeds, the other yells. Because females do not yell at such moments, this protest apparently exposes the error and the male-male tandem dissolves.)[33] The male-female pairs produce many masses of up to several dozen eggs apiece, attaching them to emergent grasses and other plants.

Spadefoot tadpoles wriggling out of these eggs face two major challenges. First, the flooded basins in which they live are very poor in food (having been dry grassland swales before the recent rain, they have little standing organic matter). Second, the water may well evaporate soon, especially in shallow and wide depressions beneath a hot prairie sun. To escape death by desiccation, the tadpoles must eat and metamorphose quickly into sturdy, terrestrial, drought-resistant toads. Not surprisingly, they waste little time in the egg, emerging in roughly two days (versus a week for species inhabiting permanent water). Their next day is spent developing mouthparts while digesting the rest of the yolk mass taken with them from the egg. Then they begin eating anything they can find, mainly algae and mud (from which they selectively digest detritus). They are especially fond of meat, quickly scavenging dead earthworms and soft-bodied insect larvae.[34]

Within a few days, the tadpoles in these ephemeral nurseries show so much variation in size and proportions that they were initially thought to belong to two different species, though they turned out to represent different developmental strategies within the same sibships. Some are small-bodied and slow-growing, while others have much larger heads (supporting grotesque serrated beaks) and grow very rapidly. Much careful work by David Pfennig has shown that these two options generate different fitness dividends, depending on local conditions and unpredictable events. One key element is the puddle's population of micro-

Amphibians that lay their eggs in ephemeral ponds and puddles face the risk of having the entire nursery evaporate before the young can metamorphose. One way of accelerating early development involves consuming puddle-mates for their nutrients. *Top:* A tiger salamander cannibal-type larva eats its sibling. In this species, cannibals have been shown to prefer unrelated victims when they are available, but often have to consume full siblings in a pinch. (Photo: David Pfennig.) *Bottom:* A cannibal-type spadefoot toad tadpole lunges at a similar-aged puddle-mate that is developing more slowly as an omnivore. (Photo: David Sanders.)

scopic crustaceans called fairy shrimp. If the puddle is reasonably stable, tadpoles consume these in small amounts along with the generalized omnivorous diet that produces slow growth, but if the puddle is shrinking rapidly its chemistry changes and a fairy shrimp "bloom" occurs. Tadpoles eating significant numbers of these shrimp then face a crossroads. According to reliable cues (water chemistry, shrimp in the diet), desiccation is an imminent threat. One solution is to shift developmental trajectories and become a fast-growing carnivore. This option has the obvious merit of improving chances for metamorphosis and survival to adulthood. The other solution is to gamble on the side of patience, betting that there will still be time to finish on the slow plan.

Both pathways have advantages and disadvantages. The carnivores have to find plenty of meat in order to graduate early, and they resort to considerable amounts of cannibalism. If they are eating truly doomed siblings, the sacrifice in inclusive fitness is probably trivial. But if conditions do turn out to be sufficiently stable to support slow development, they are eating viable kin. Furthermore, the cannibal tadpoles' haste in completing metamorphosis leaves them with smaller bodies as adults. And being small has delayed costs, since undersized females produce fewer eggs and undersized males are more easily pried off females when mating.[35]

This is where the ability to recognize kin might come in handy: being able to choose victims on the basis of relatedness might confer a fitness advantage if coupled with cannibalism. A cannibal that could avoid its own siblings while consuming unrelated puddle-mates would not have to sacrifice indirect fitness (via kin) in order to promote its own future. Pfennig pursued this idea with lab experiments and found that spadefoot tadpoles of the large-jawed "cannibal" type prefer to hang out with non-siblings, but those of the small-bodied "omnivore" type tend to congregate with their sibs. Two other provocative clues emerged. First, cannibals seem to nip at potential victims before seizing them (possibly a taste test?). Second, their choice of close kin seems to wane when they are less hungry, plausibly indicating that sib-avoidance is a luxury that might well be dropped in emergencies.[36]

These questions have also been addressed with another desert-dwelling amphibian, the tiger salamander, which has a generally similar method for escaping from evaporating desert pools. The likelihood that salamander larvae will switch to cannibal types increases when the nursery is crowded, especially if the crowd includes some non-siblings.[37] The cannibals prefer to eat non-kin in general, but make their mealtime decision more rapidly when the choice of victims is between an unrelated tadpole and a sibling than when it is between an unrelated tadpole and a cousin. Their ability to show such discrimination is lost if the nose is plugged, implicating sense of smell as playing a vital role.[38]

This fascinating research is continuing with explorations of some new costs for eating kin. It is coming to light that closely related cannibal victims can harbor parasites and other disease organisms that are especially harmful to close genetic relatives, making kin an especially dangerous menu item.[39] That finding suggests that recognizing and avoiding kin during acts of cannibalism may be favored by natural selection because of its impact on the cannibal's personal fitness (the direct component), but the role of kin selection (indirect fitness) in shaping such behavior also continues to be supported strongly.[40] The mix of selfishness and nepotism is unlikely to be simple in real-world systems.

Epilogue

As an adolescent I aspired to lasting fame, I craved factual certainty, and I thirsted for a meaningful vision of human life—so I became a scientist. This is like becoming an archbishop so you can meet girls.

—Matt Cartmill

In Greco-Roman mythology, the Corinthian king Sisyphus was condemned to roll a great stone up a steep slope in Hades, only to have it roll back down by itself, requiring him to repeat his labor endlessly. When I was a college sophomore in the 1960s, this is pretty much how biological research first appeared to me: one generation's findings are published and deposited in the university library, thereby forcing members of the next generation to labor their way through the growing number of pages until they, too, publish and condemn a third generation to drudgery, and so on. It did not occur to me that there was any excitement, stimulation, or fun involved.

Even so, like a moth to light I was drawn to biology, especially the beauty of adaptation. I marveled at how wonderfully organisms compensate for limitations and how exquisitely the disparate pieces of nature fit together. Snakes make up for their lack of appendages (and leverage for dismembering prey) by having jaws that disengage to allow them to swallow oversized items whole. Lizards, insects, and mice carefully choose to live near particular plants and other substrates whose colors and patterns they match exactly. Bats use ultrasonic vocalizations to generate echoes that allow them to navigate in the dark. While amateur naturalists had meticulously described some of these facts, many of the newer discoveries could not possibly have been detected through casual observation. For example, to discover that Pacific salmon return after years at sea to their precise natal streams before spawning and dying required that somebody tag colossal numbers of fry and then sample the adults diligently. To reveal what birdsong means to birds similarly takes astonishing amounts of effort and experimentation, not to mention the invention of portable tape recorders.

Like many other budding biologists of the era, I grew up on TV programs about nature without any thought to how the producers and film crews could possibly know about the phenomena they were relating. It never dawned on me that such programs had to be based on the detailed studies of scientists. But at some point I must have realized that research is not a Sisyphean nightmare at all but a kaleidoscopic process. During the latter half of my undergraduate years I began to meet active researchers who were excited about their work, and I learned that the rock never rolls back to the same spot. Sometimes it doesn't roll back down at all, but even when it does it often comes to rest in a more interesting location. With that enlightenment came the corollary that papers were written, published, and deposited in libraries for reasons other than to burden undergraduates.

Over the decades since then, I've learned many unexpected things about field research itself, in addition to the mysteries explored in these pages. For example, I realized that the moments of great exhilaration and discovery are rare outposts in vast deserts of labor and tedium. Neither the TV shows nor the fifteen-minute oral presentations at conferences show all the dead ends and frustrations, for reasons that now seem obvious.

One day in the late 1990s I tapped on the door of my colleague Mike Kaspari for a few moments of chitchat and was greeted with a surprising question. At what point in my career, Mike wanted to know, had I realized that it's much more exciting when the data come out exactly the opposite of what you expected? I remember frowning slightly as I thought about this, then replying "I think you've just cleared up something that's been bothering me for a long time."

The process of science consists of making proposals and then testing them to see which ones provide the best explanations. In a world of increasing specialization across all fields, it is hardly surprising that biology has become increasingly split between those who focus mainly on generating the candidate ideas (theoreticians) and those who subject them to tests (empiricists). Empiricism, of course, is not a new development; the recording of facts from natural history goes back to the invention of writing (or hieroglyphics), and the oral traditions extend well before that. The label derives from the Empirics, a group of thirteenth-century British physicians/healers who roundly rejected the whole idea of generalizations and conjecture, preferring to deal only with the concrete substance of the Facts. This extreme position must not have taken them very far in the pursuit of cures, but the term caught on. Nowadays, biologists are consigned to the empirical end of the scientific spectrum if they devote research time mainly to the collection and analysis of data (that is, numbers). At the other end of that spectrum lies an often-intimidating jungle of mathematical constructs, wherein the numbers are allowed to play more freely. When the duality works best, the models are built to address problems rooted in the rich soil of fact and they offer testable possibilities for how nature might work. To the extent that they contain components one can measure, they can suggest what we most need to know next.

I try to be a modern empiricist by paying attention to what the theoreticians propose but then letting the animals make the call. A hunch that has been formalized as a mathematical model is still just a hunch. As described in these chapters, many of the things I most clearly expected to see simply did not appear. I confidently anticipated that egret parents would stop the siblicidal fights: they did not. I was sure that egret nestlings adjust their aggressiveness in proportion to their hunger: they do not. But some of the guiding predictions were supported. The switch by great blue heron nestlings from pacifism to lethal aggressiveness, both when sent to live with egret foster parents and in the heron colony in Quebec, fit my hunch about how the size of prey affects the value of fighting. And the cattle egrets did cut way down on aggression after brood reduction, whether real or simulated. Sometimes one is on the right track; other times, not.

Being wrong has one great virtue, especially with complex puzzles: it forces you to think harder. As when you search for misplaced car keys, once you've checked all the obvious places you have to start imagining less probable scenarios. In scientific research, finding the expected can produce a false sense of having solved the problem and cause you to relax. You're less motivated to keep searching and see what else might turn up. This is what Mike Kaspari was getting at. When the simple explanation does not fit, you have to keep picking at the problem until you discern a different possible solution. And then you test that one.

It is natural for us to prefer simple, easy-to-grasp explanations. Even more seductive are those that add the element of novelty. When a new hypothesis emerges in my field it often seems as if some empiricists gravitate to the role of cheerleader, rather than that of rigorous tester. This can lead to the fast collecting of inconclusive data that are merely "consistent with" the idea. Many journal reviewers like a tidy story, so such results are easier to publish. Unfortunately, durable science usually has to be done the hard way. As Einstein put it, "No amount of experimentation can prove me right. A single experiment can prove me wrong."

The mix of theory and data put together in this book is up to date at the time of its writing, but much is doomed to eventual obsolescence. Scientific knowledge is nothing if not ephemeral, and only nonscientists misinterpret that as a flaw.

Notes

1. In a Family Way

1. Hamilton 1964a, 1964b.
2. Cramer and Blass 1983.
3. Trillmich 1986.
4. Carl 1987.

2. The Problem with Sex

1. Ahnesjö 1996.
2. Whether or not Haldane made the quip about "two brothers or eight cousins," it is clear that he understood the underlying point. He wrote in a paper published in 1955 (p. 44): "Let us suppose that you carry a rare gene which affects your behaviour so that you jump into a flooded river and save a child, but you have one chance in ten of being drowned, while I do not possess the gene, and stand on the bank and watch the child drown. If the child is your own child or your brother or sister, there is an even chance that the child will also have this gene, so five such genes will be saved in children for one lost in an adult. If you save a grandchild or nephew the advantage is only two and a half to one."
3. Sherman 1977.

3. Nursery Life with Attitude

1. Hart 1973.
2. Gilmore, Dodrill, and Linley 1983.
3. Springer 1948.
4. Nuechterlein 1981.

4. The Trouble with Parents

1. Parker, Mock, and Lamey 1989.
2. Blick 1977, Parker and Macnair 1978, Stamps, Metcalf, and Krishnan 1978.
3. From Stamps and Metcalf 1980.
4. Mock and Forbes 1995, Mock and Parker 1997.
5. Temme and Charnov 1987.
6. Lyon, Eadie, and Hamilton 1994.
7. Bortolotti, Wiebe, and Iko 1991.

5. Raising Cain

1. Rowe 1947, Gargett 1978. For a summary of essentially everything known about black eagles, see Gargett 1990.
2. Cash and Evans 1986, Evans and Macmahon 1987.
3. Ingram 1959, 1962.
4. For an excellent collection of review chapters on cannibalism across all forms of animals, see Elgar and Crespi 1992.
5. Stinson 1979.
6. Brown, Gargett, and Steyn 1977, p. 79.
7. Ibid., p. 70.
8. Forbes and Mock 2000, Mock, Drummond, and Stinson 1990.
9. Gargett 1977.
10. Mock 1984a.
11. Cash and Evans 1986, Evans and Macmahon 1987.
12. Davies 1992.
13. Forbes 1990.
14. Edwards and Collopy 1983, Bortolotti 1986a, Anderson 1989.
15. Ploger 1997.
16. Högstedt 1980.
17. Simmons 1997.
18. Gargett 1990.
19. Johnston, Bednarz, and Zack 1987.
20. Griffin, Paton, and Baskett 1998.
21. See Forbes 1991.
22. Lack 1968.
23. Monaghan, Nager, and Houston 1998.
24. Heaney and Monaghan 1996.

6. Killing Me Softly

1. Ganeshaiah and Uma Shaanker 1988.
2. O'Gara 1969.
3. Weir 1971a, 1971b.
4. Birney and Baird 1985.
5. Boersma and Stokes 1995.
6. Slagsvold et al. 1984.
7. Lamey 1992, St. Clair 1992.
8. Lamey 1993, St. Clair 1996.
9. Lamey 1992.
10. St. Clair et al. 1995.
11. Gould and Lewontin 1979.
12. St. Clair et al. 1995.

7. Parenting in an Uncertain World

1. Lack 1947.
2. Stanback and Koenig 1992.
3. Reynolds 1996.
4. Ricklefs 1965.
5. Vince 1969, but see also Schwagmeyer et al. 1991.
6. Sutton 1967.

8. The Ultimate Food Fight

1. Stoleson and Beissinger 1995.
2. Mock and Lamey 1991.
3. Mock and Parker 1986; method amended by Lamey, Evans, and Hunt 1996.
4. Forbes and Ydenberg 1992.
5. Mock and Parker 1986.
6. Mock 1985.
7. Mock 1984b.
8. Mock, Lamey, and Ploger 1987.
9. Lack 1966, p. 309 (italics added).
10. Braun and Hunt 1983.
11. Poole 1982.
12. Drummond and Osorno 1992.
13. Drummond, Gonzalez, and Osorno 1986, Drummond et al. 1991.

14. Drummond and García Chavelas 1989.
15. Machmer and Ydenberg 1998.
16. Cook, Monoghan, and Burns 2000.
17. For a completely different interpretation, see Creighton and Schnell 1996, Drummond 2001.
18. Mock and Lamey 1991.

9. Gambling with Children

1. Reviewed in Andersson 1994, Searcy and Yasukawa 1995, Beletsky 1996.
2. Beletsky 1996.
3. Emlen and Oring 1977.
4. Beletsky 1996.
5. Searcy and Yasukawa 1995.
6. Bray, Kennelly, and Guarino 1975.
7. Gibbs et al. 1990, Westneat 1993, Gray 1997.
8. Gibbs et al. 1990, Westneat 1993, Gray 1997.
9. Calculated as $r = p \times .5 + (1 - p) \times .25$, where $p = 3/4$ and $(1 - p) = 1/4$. Thus $r = .4375$.
10. Searcy and Yasukawa 1995.
11. Forbes et al. 1997.
12. Clark and Wilson 1981, Stoleson and Beissinger 1995.
13. Lack 1947, 1948.
14. Davidson and Andrewartha 1948a, 1948b.
15. Andrewartha and Birch 1954.
16. Ricklefs 1990.
17. Lamey and Lamey 1994.
18. Lack 1954.
19. Wynne-Edwards 1962.
20. Ydenberg and Bertram 1989.
21. Reviewed in Lack 1966.
22. See reviews in Clark and Wilson 1981, Amundsen and Stokland 1988.
23. Vermeer 1963.
24. Nelson 1964.
25. Reid 1987.
26. Summarized in Ydenberg and Bertram 1989.
27. Jarvis 1974.
28. Williams 1966b, Smith and Fretwell 1968, Charnov and Krebs 1974.
29. Fisher 1930.

30. Reid 1987.
31. Heaney and Monaghan 1996, Monaghan, Nager, and Houston 1998.
32. Magrath 1989.
33. Forbes and Glassey 2000.

10. Beggars, Cheats, and Bad Fruit

1. Dawkins and Krebs 1978.
2. Rodríguez-Gironés 1996.
3. Haskell 1994. For a review see Haskell 2002.
4. Lloyd 1986.
5. Trivers 1974.
6. See Stamps 1993, Kilner and Johnstone 1997. For reviews of work on avian begging, see Wright and Leonard 2002.
7. McRae, Weatherhead, and Montgomerie 1993.
8. Smith and Montgomerie 1991.
9. Kacelnik et al. 1995.
10. Roulin 2002.
11. Glassey and Forbes 2002.
12. Price, Harvey, and Ydenberg 1996.
13. Price and Ydenberg 1995.
14. Stamps et al. 1989.
15. Gottlander 1987, Lyon, Eadie, and Hamilton 1994, Leonard and Horn 1996.
16. Boersma 1991, Boersma and Stokes 1995.
17. Kilner, Noble, and Davies 1999.
18. For supportive reviews of honest signal theory, see Kilner and Johnstone 1997, Johnstone and Godfray 2002; for a summary of some of its problems, see Royle, Hartley, and Parker 2002.
19. For a highly readable account of cuckoo parasitism, see Davies 2000.
20. Lotem 1993.
21. Davies, Kilner, and Noble 1998, Kilner, Noble, and Davies 1999.
22. Weatherhead 1989, Payne 1997.
23. Lichtenstein 1998.
24. Briskie and Sealy 1987.
25. Sealy 1995.
26. Lichtenstein and Sealy 1998.
27. Sato 1986.
28. Davies and Brooke 1989.

29. Eadie and Lyon 1998.
30. Hamilton 1964b, p. 49.
31. Haig and Graham 1991.
32. Haig and Westoby 1991.

11. Silly Squabbles and Serious Sabotage

1. Trivers 1974.
2. Fagerström 1987.
3. Dawkins 1989, p. 298.
4. Reviewed in Mock and Parker 1997.
5. Godfray 1991, 1995a, 1995b, Johnstone 1997, Kilner and Johnstone 1997, Kilner, Noble, and Davies 1999, Wright and Leonard 2002.
6. Ploger 1997.
7. Blaker 1969, Jenni 1969, Werschkul 1979.
8. Forbes and Mock 1996.
9. Drummond, Gonzalez, and Osorno 1986, Mock 1987.
10. Hahn 1981.
11. Werschkul 1979.
12. Ploger and Mock 1986.
13. Fujioka 1985.
14. Mock and Ploger 1987.
15. Maynard Smith and Parker 1976.
16. Osorno and Drummond 1995.
17. Hébert and Barclay 1986 (gull), Gibbons 1987 (jackdaw), Hébert and Sealy 1993 (warbler), Wiebe and Bortolotti 1994b (kestrel), Machmer and Ydenberg 1998 (osprey).
18. Wiebe and Bortolotti 1994a.
19. Bertram 1979.
20. Vehrencamp 1977, Macedo, Cariello, and Muniz 2001.
21. Vesey-Fitzgerald 1957, Nuechterlein and Johnson 1981, Horsfall 1984.
22. Howe 1976, see review in Williams 1994.
23. Lamey 1990, Slagsvold et al. 1984.
24. Tortosa and Redondo 1992, Kłosowski, Kłosowski, and Zieliński 2002.
25. Bustamente, Cuervo, and Moreno 1992, van Heezik and Seddon 1996, Bustamente, Boersma, and Davis 2002.
26. Young 1963.
27. Gerrard and Bortolotti 1988.
28. Viñuela 1999.
29. Lougheed and Anderson 1999.

12. Parent-Offspring Conflict Revisited

1. Ono et al. 1995.
2. Withers, Fahrbach, and Robinson 1995.
3. For any number n of male partners, the coefficient of relatedness for two random daughters is $.75 \times 1/n + .25 \times [(n - 1)/n]$, the weighted average of whether they're full or half-sisters.
4. Strassmann et al. 1989.
5. Ratniecks 1988, Ratniecks and Visscher 1989.
6. Visscher 1996.
7. For a review see Godfray 1994.
8. Cruz 1981, 1986.
9. Trivers and Hare 1976.
10. Harvey, Corley, and Strand 2000.
11. Grbíc, Ode, and Strand 1992.
12. Mock and Forbes 1992.
13. LeMasurier 1987, Godfray 1994.
14. Hamilton 1967, 1979.
15. Alexander and Sherman 1977.
16. Nonacs 1986, Bull and Charnov 1988, Clutton-Brock 1991, Seger 1991.
17. Boomsma 1991.
18. Mueller 1991.

13. Till Death Do Us Part

1. Parker 1985.
2. Lack 1968.
3. Rees et al. 1996.
4. Black 1996.
5. Desrochers and Magrath 1996.
6. Cézilly and Johnson 1995.
7. Maynard Smith 1982.
8. Maynard Smith 1977, Chase 1980, Houston and Davies 1985, Parker 1985, Winkler 1987, Winkler and Wallin 1987, Hussell 1988.
9. Chase 1980, Houston and Davies 1985.
10. von Haartman 1953.
11. Drent and Daan 1980.
12. Wright and Cuthill 1989.
13. Rohwer 1978; evidence reviewed in FitzGerald and Whoriskey 1992, Fitzgerald 1992; theoretical summary in Sargent 1992.

14. Meffe and Crump 1987.
15. Gustafsson and Sutherland 1988.
16. O'Connor 1978.
17. Mock and Parker 1997, 1998.
18. Spear and Nur 1994.
19. Mills, Yarrall, and Mills 1996.
20. Inoue 1985.
21. Thomas, Kayser, and Hafner 1999.
22. Clutton-Brock and Parker 1995.
23. Rohwer 1978.
24. Dawkins and Carlisle 1976.
25. Trivers 1972.
26. Nakamura and Kramer 1982.
27. Smith 1979a, 1979b.
28. Beissinger and Snyder 1987.
29. Beissinger 1987.
30. Beissinger and Snyder 1987.

14. Upgrading the Kids

1. Thornhill 1976, reviewed in Thornhill and Alcock 1983.
2. Gwynne 1981, Gwynne and Simmons 1990.
3. Andrade 1996.
4. Petrie, Halliday, and Sanders 1991, Petrie 1994.
5. Buchholz 1922.
6. Stephenson 1981, Stephenson and Windsor 1986, Stearns 1987, Kozlowski and Stearns 1989.
7. Willson and Burley 1983, Queller 1989.
8. Searcy and Macnair 1993.
9. Stephenson 1981.
10. Kahl 1964.
11. Tyndale-Biscoe and Renfree 1987, Cockburn 1989.
12. Tait 1980.
13. Komdeur et al. 1997.
14. Voltura 1998.
15. Legge 2000.
16. Edwards and Collopy 1983, Bortolotti 1986a, Bortolotti 1986b, Olsen and Cockburn 1991, Drummond et al. 1991.
17. Charnov et al. 1981.
18. Hamilton and Zuk 1982.

19. Zuk et al. 1990, Ligon et al. 1990.
20. Kilner 1997.
21. Lyon, Eadie, and Hamilton 1994.
22. Horsfall 1984.
23. Forbes and Mock 1998.
24. Simmons 1988.
25. Forbes and Mock 1998.
26. Haig 1987, 1992.
27. Forbes and Mock 1998.

15. Together Again

1. Female mouse and hamster pups that were sandwiched *in utero* between two brothers are more aggressive and tend to have many more sons than daughters. See vom Saal and Bronson 1980, vom Saal 1984, Clark, Tucker, and Galef 1992, Clark, Karpluk, and Galef 1993.
2. As has been reported for pronghorns; O'Gara 1969.
3. Teicher and Blass 1976, 1977.
4. Fraser 1990.
5. Fraser and Thompson 1990.
6. Fraser 1990.
7. Bryant and Tatner 1990.
8. Frank, Glickman, and Licht 1991, Golla, Hofer, and East 1999.
9. Frank, Glickman, and Licht 1991, Smale et al. 1995.
10. Golla, Hofer, and East 1999.
11. Legge and Cockburn 2000.
12. Legge 2000, Nathan, Legge, and Cockburn 2001.
13. Hrdy 1977, 1979.
14. Stanback and Koenig 1992.
15. Parsons 1971.
16. Gould 1982.
17. Anderson 1991.
18. Drummond, Gonzalez, and Osorno 1986.
19. Urrutia and Drummond 1990.
20. Creighton 1995.
21. Bartlett 1987.
22. Reviewed in Elwood 1992.
23. Schwagmeyer 1979, Huck 1984, Elwood 1992.
24. Berger 1983.
25. Bruce 1959, Bruce and Parrott 1960.

26. Hausfater and Hrdy 1994.
27. Kłosowski, Kłosowski, and Zieliński 2002.
28. For a summary of the project see Koenig and Mumme 1987.
29. Mumme, Koenig, and Pitelka 1983.
30. Packer and Pusey 1983a, 1983b.
31. Waldman and Adler 1979; see reviews by Waldman 1988, Blaustein and Waldman 1992, Sherman, Reeve, and Pfennig 1997.
32. Waldman 1988.
33. Bragg 1965.
34. Bragg 1965.
35. Pfennig 1992.
36. Pfennig, Reeve, and Sherman 1993.
37. Pfennig and Collins 1993.
38. Pfennig, Sherman, and Collins 1994.
39. Pfennig, Ho, and Hoffman 1998.
40. Pfennig, Collins, and Ziemba 1999.

Works Cited

Ahnesjö, I. 1996. Apparent resource competition among embryos in the brood patch of a male pipefish. *Behavioral Ecology and Sociobiology* 38: 167–172.

Alexander, R. D., and P. W. Sherman. 1977. Local mate competition and parental investment in social insects. *Science* 196: 494–500.

Amundsen, T., and J. N. Stokland. 1988. Adaptive significance of asynchronous hatching in the shag: A test of the brood reduction hypothesis. *Journal of Animal Ecology* 57: 329–344.

Anderson, D. J. 1989. The role of hatching asynchrony in siblicidal brood reduction of two booby species. *Behavioral Ecology and Sociobiology* 25: 363–368.

——— 1991. Parent blue-footed boobies are not infanticidal. *Ornis Scandinavica* 22: 169–170.

Andersson, M. 1994. *Sexual selection.* Princeton University Press.

Andewartha, H. G., and L. C. Birch. 1954. *The distribution and abundance of animals.* University of Chicago Press.

Andrade, M. C. B. 1996. Sexual selection for male sacrifice in the Australian redback spider. *Science* 271: 70–72.

Bartlett, J. 1987. Filial cannibalism in burying beetles. *Behavioral Ecology and Sociobiology* 21: 179–183.

Beissinger, S. R. 1987. Mate desertion and reproductive effort in the snail kite. *Animal Behaviour* 35: 1504–19.

Beissinger, S. R., and N. F. R. Snyder. 1987. Mate desertion in the snail kite. *Animal Behaviour* 35: 477–487.

Beletsky, L. 1996. *The red-winged blackbird: The biology of a strongly polygynous songbird.* Academic Press.

Berger, J. L. 1983. Induced abortion and social factors in wild horses. *Nature* 303: 59–61.

Bertram, B. C. R. 1979. Ostriches recognize their own eggs and discard others. *Nature* 279: 233–234.

Birney, E. C., and D. D. Baird. 1985. Why do some mammals polyovulate to produce a litter of two? *American Naturalist* 126: 136–140.

Black, J. M., ed. 1996. *Partnerships in birds: The study of monogamy.* Oxford University Press.

Blaker, D. 1969. Behaviour of the cattle egret *Ardeola ibis. Ostrich* 40: 75–129.

Blaustein, A. R., and B. Waldman. 1992. Kin recognition in anuran amphibians. *Animal Behaviour* 44: 207–222.

Blick, J. E. 1977. Selection for traits which lower individual reproduction. *Journal of Theoretical Biology* 67: 597–601.

Boersma, P. D. 1991. Asynchronous hatching and food allocation in the Magellanic penguin *Spheniscus magellanicus. Acta XX Congressus Internationalis Ornithologicus* II: 961–973.

Boersma, P. D., and P. D. Stokes. 1995. Mortality patterns, hatching asynchrony, and size asymmetry in Magellanic penguin *(Spheniscus magellanicus)* chicks. In P. Dann and P. Reilly, eds., *Penguin biology 2,* 15–43. Surrey Beatty and Sons.

Boomsma, J. J. 1991. Adaptive colony sex ratios in primitively eusocial bees. *Trends in Ecology and Evolution* 6: 92–95.

Bortolotti, G. R. 1986a. Evolution of growth rates in eagles: Sibling competition vs. energy considerations. *Ecology* 67: 182–194.

——— 1986b. Influence of sibling competition on nestling sex ratios of sexually dimorphic birds. *American Naturalist* 127: 495–507.

Bortolotti, G. R., K. L. Wiebe, and W. M. Iko. 1991. Cannibalism of nestling American kestrels by their parents and siblings. *Canadian Journal of Zoology* 69: 1447–53.

Bragg, A. N. 1965. *Gnomes of the night.* University of Pennsylvania Press.

Braun, B. M., and G. L. Hunt Jr. 1983. Brood reduction in black-legged kittiwakes. *Auk* 100: 469–476.

Bray, O. E., J. J. Kennelly, and J. L. Guarino. 1975. Fertility of eggs produced on territories of vasectomized red-winged blackbirds. *Wilson Bulletin* 87: 187–195.

Briskie, J., and S. G. Sealy. 1987. Responses of least flycatchers to experimental inter- and intraspecific brood parasitism. *Condor* 89: 899–901.

Brown, L. H., V. Gargett, and P. Steyn. 1977. Breeding success in some African eagles relative to theories about sibling aggression and its effects. *Ostrich* 48: 65–71.

Bruce, H. M. 1959. An exteroceptive block to pregnancy in the mouse. *Nature* 184: 105.

Bruce, H. M., and D. M. V. Parrott. 1960. Role of olfactory sense in pregnancy block by strange males. *Science* 131: 1526.

Bryant, D. M., and P. Tatner. 1990. Hatching asynchrony, sibling competition and siblicide in nestling birds: Studies of swiftlets and bee-eaters. *Animal Behaviour* 39: 657–671.

Buchholz, J. T. 1922. Developmental selection in vascular plants. *Botanical Gazette* 73: 249–286.

Bull, J. J., and E. L. Charnov. 1988. How fundamental are Fisherian sex ratios? In P. H. Harvey and L. Partridge, eds., *Oxford surveys in evolutionary biology,* vol. 5, 96–135. Oxford University Press.

Bustamente, J., P. D. Boersma, and L. S. Davis. 2002. Feeding chases in penguins: Begging competition on the run? In J. Wright and M. L. Leonard, eds., *Begging in birds,* 303–318. Kluwer Academic.

Bustamente, J., J. J. Cuervo, and J. Moreno. 1992. The function of feeding chases in the chinstrap penguin *Pygoscelis antarctica. Animal Behaviour* 44: 753–759.

Carl, R. A. 1987. Age-class variation in foraging techniques by brown pelicans. *Condor* 89: 525–533.

Cash, K., and R. M. Evans. 1986. Brood reduction in the American white pelican *(Pelecanus erythrorhynchos). Behavioral Ecology and Sociobiology* 18: 413–418.

Cézilly, F., and A. R. Johnson. 1995. Re-mating between and within breeding seasons in the Greater Flamingo *Phoenicopterus ruber roseus. Ibis* 137: 543–546.

Charnov, E. L. 1982. *The theory of sex allocation.* Princeton University Press.

Charnov, E. L., and J. Krebs. 1974. On clutch-size and fitness. *Ibis* 116: 217–219.

Charnov, E. L., R. L. Los-Den Hartogh, W. T. Jones, and J. Van Den Assem. 1981. Sex ratio evolution in a variable environment. *Nature* 289: 27–33.

Chase, I. D. 1980. Cooperative and noncooperative behavior in animals. *American Naturalist* 115: 827–857.

Clark, A. B., and D. S. Wilson. 1981. Avian breeding adaptations: Hatching asynchrony, brood reduction and nest failure. *Quarterly Review of Biology* 56: 253–277.

Clark, M. M., P. Karpluk, and B. G. Galef Jr. 1993. Hormonally mediated inheritance of acquired characteristics in Mongolian gerbils. *Nature* 364: 712.

Clark, M. M., L. Tucker, and B. G. Galef Jr. 1992. Stud males and dud males: Intra-uterine position effects on the reproductive success of male gerbils. *Animal Behaviour* 43: 215–221.

Clutton-Brock, T. H. 1991. *The evolution of parental care.* Princeton University Press.

Clutton-Brock, T. H., and G. A. Parker. 1995. Punishment in animal societies. *Nature* 373: 209–216.

Cockburn, A. 1989. Adaptive patterns in marsupial reproduction. *Trends in Ecology and Evolution* 4: 126–130.

Cook, M. I., P. Monoghan, and M. D. Burns. 2000. Effects of short-term hunger and competitive asymmetry on facultative aggression in nestling black guillemots *Cepphus grylle*. *Behavioral Ecology* 11: 282–297.

Cramer, C. P., and E. M. Blass. 1983. Mechanisms of control of milk intake in suckling rats. *American Journal of Physiology* 245: R154–159.

Creighton, J. C. 1995. Factors affecting brood size in the burying beetle *Nicrophorus orbicollis*. Ph.D. diss., University of Oklahoma, Norman.

Creighton, J. C., and G. D. Schnell. 1996. Proximate control of siblicide in cattle egrets: A test of the food amount hypothesis. *Behavioral Ecology and Sociobiology* 38: 371–378.

Cruz, Y. P. 1981. A sterile defender morph in a polyembryonic hymenopterous parasite. *Nature* 294: 446–447.

——— 1986. The defender role of the precocious larvae of *Copidomopsis tanytmenus* Caltagirone (Encyrtidae, Hymenoptera). *Journal of Experimental Zoology* 237: 309–318.

Darwin, C. 1859. *On the origin of species*. John Murray.

Davidson, J., and H. G. Andrewartha. 1948a. Annual trends in a natural population of *Thrips imaginis* (Thrysanoptera). *Journal of Animal Ecology* 17: 193–199.

——— 1948b. The influence of rainfall, evaporation and atmospheric temperature on fluctuations in the size of a natural population of *Thrips imaginis* (Thrysanoptera). *Journal of Animal Ecology* 17: 200–222.

Davies, N. B. 1992. *Dunnock behaviour and social evolution*. Oxford University Press.

——— 2000. *Cuckoos, cowbirds and other cheats*. Academic Press.

Davies, N. B., and M. d. L. Brooke. 1989. Cuckoos and parasitic ants: Interespecific brood parasitism as an evolutionary arms race. *Trends in Ecology and Evolution* 4: 274–278.

Davies, N. B., R. M. Kilner, and D. G. Noble. 1998. Nestling cuckoos, *Cuculus canorus*, exploit hosts with begging calls that mimic a brood. *Proceedings of the Royal Society of London B* 265: 673–678.

Dawkins, R. 1989. *The selfish gene*. New ed. Oxford University Press.

Dawkins, R., and T. R. Carlisle. 1976. Mate desertion, parental investment, and a fallacy. *Nature* 262: 131–132.

Dawkins, R., and J. R. Krebs. 1978. Animal signals. In J. R. Krebs and N. B. Davies, eds., *Behavioural ecology: An evolutionary approach*, 1st ed., 282–309. Sinauer.

Desrochers, A., and R. D. Magrath. 1996. Divorce in the European blackbird: Seeking greener pastures? In J. M. Black, ed., *Partnerships in birds,* 177–191. Oxford University Press.

Drent, R. H., and S. Daan. 1980. The prudent parent: Energetic adjustments in avian breeding. *Ardea* 68: 225–252.

Drummond, H. 2001. A re-evaluation of the role of food in nestling aggression. *Animal Behaviour* 61: 1–10.

Drummond, H., and C. Garcia Chavelas. 1989. Food shortage influences sibling aggression in the blue-footed booby. *Animal Behaviour* 37: 806–819.

Drummond, H., E. Gonzalez, and J. L. Osorno. 1986. Parent-offspring co-operation in the blue-footed booby, *Sula nebouxii. Behavioral Ecology and Sociobiology* 19: 365–392.

Drummond, H., and J. L. Osorno. 1992. Training siblings to be submissive losers: Dominance between booby nestlings. *Animal Behaviour* 44: 881–893.

Drummond, H., J. L. Osorno, R. Torres, C. Garcia Chavelas, and H. M. Larios. 1991. Sexual size dimorphism and sibling competition: Implications for avian sex ratios. *American Naturalist* 138: 623–641.

Eadie, J. M., and B. E. Lyon. 1998. Cooperation, conflict, and crèching behavior in goldeneye ducks. *American Naturalist* 151: 397–408.

Edwards, T. C. Jr., and M. W. Collopy. 1983. Obligate and facultative brood reduction in eagles: An examination of factors that influence fratricide. *Auk* 100: 630–635.

Elgar, M. A., and B. J. Crespi, eds. 1992. *Cannibalism: Ecology and evolution among diverse taxa.* Oxford University Press.

Elwood, R. W. 1992. Pup-cannibalism in rodents: Causes and consequences. In M. A. Elgar and B. J. Crespi, eds., *Cannibalism,* 299–322. Oxford University Press.

Emlen, S. T., and L. W. Oring. 1977. Ecology, sexual selection, and the evolution of mating systems. *Science* 197: 215–223.

Evans, R. M., and B. F. MacMahon. 1987. Within-brood variation growth and conditions in relation to brood reduction in the American white pelican. *Wilson Bulletin* 99: 190–201.

Fagerström, T. 1987. On theory, data, and mathematics in ecology. *Oikos* 50: 258–261.

Fisher, R. A. 1930. *The genetical theory of natural selection.* Clarendon Press.

FitzGerald, G. J. 1992. Filial cannibalism in fishes: Why do parents eat their offspring? *Trends in Ecology and Evolution* 7: 7–10.

FitzGerald, G. J., and F. G. Whoriskey. 1992. Empirical studies of cannibalism in fish. In M. A. Elgar and B. J. Crespi, eds., *Cannibalism,* 228–255. Oxford University Press.

Forbes, L. S. 1990. Insurance offspring and the evolution of avian clutch size. *Journal of Theoretical Biology* 147: 345–359.
——— 1991. Optimal offspring size and number in a variable environment. *Journal of Theoretical Biology* 150: 299–304.
Forbes, L. S., and B. Glassey. 2000. Asymmetric sibling rivalry in red-winged blackbirds. *Behavioral Ecology and Sociobiology* 48: 413–417.
Forbes, L. S., and D. W. Mock. 1996. Food, information, and avian brood reduction. *Écoscience* 3: 45–53.
——— 1998. Parental optimism and progeny choice: When is screening for offspring quality affordable? *Journal of Theoretical Biology* 192: 3–14.
——— 2000. A tale of two strategies: Life-history aspects of family strife. *Condor* 102: 23–34.
Forbes, L. S., S. Thornton, B. Glassey, M. Forbes, and N. J. Buckley. 1997. Why parent birds play favourites. *Nature* 390: 351–352.
Forbes, L. S., and R. C. Ydenberg. 1992. Sibling rivalry in a variable environment. *Theoretical Population Biology* 41: 335–360.
Frank, L. G., S. E. Glickman, and P. Licht. 1991. Fatal sibling aggression, precocial development, and androgens in neonatal spotted hyenas. *Science* 252: 702–704.
Fraser, D. 1990. Behavioural perspectives on piglet survival. *Journal of Reproduction and Fertility,* suppl. 40: 355–370.
Fraser, D., and B. K. Thompson. 1990. Armed sibling rivalry among piglets. *Behavioral Ecology and Sociobiology* 29: 9–15.
Fujioka, M. 1985. Food delivery and sibling competition in experimentally even-aged broods of the cattle egret. *Behavioral Ecology and Sociobiology* 17: 67–74.
Ganeshaiah, K. N., and R. Uma Shaanker. 1988. Seed abortion in wind-dispersed pods of *Dalbergia sissoo:* Maternal regulation or sibling rivalry? *Oecologia* 75: 135–139.
Gargett, V. 1977. A 13-year population study of the black eagles in the Matopos, Rhodesia, 1964–1976. *Ostrich* 48: 17–27.
——— 1978. Sibling aggression in the black eagle in the Matopos, Rhodesia. *Ostrich* 49: 57–63.
——— 1990. *The black eagle.* Acorn Books.
Gerrard, J. M., and G. R. Bortolotti. 1988. *The bald eagle.* Smithsonian Institution Press.
Gibbons, D. 1987. Hatching asynchrony reduces parental investment in the jackdaw. *Journal of Animal Ecology* 56: 403–414.
Gibbs, H. L., P. J. Weatherhead, P. T. Boag, B. N. White, L. M. Tabak, and D. J. Hoysak. 1990. Realized reproductive success of polygynous red-winged blackbirds revealed by DNA markers. *Science* 250: 1394–97.
Gilmore, R. G., J. W. Dodrill, and P. A. Linley. 1983. Reproduction and develop-

ment of the sand tiger shark, *Odontapsis taurus* (Rafinesque). *Fishery Bulletin* 81: 201–225.
Glassey, B., and L. S. Forbes. 2002. Muting individual nestlings reduces parental foraging for the brood. *Animal Behaviour* 63: 779–786.
Godfray, H. C. J. 1991. The signalling of need by offspring to their parents. *Nature* 353: 328–330.
——— 1994. *Parasitoids: Behavioral and evolutionary ecology.* Princeton University Press.
——— 1995a. Signalling of need between parents and young: Parent-offspring conflict and sibling rivalry. *American Naturalist* 146: 1–24.
——— 1995b. Evolutionary theory of parent-offspring conflict. *Nature* 376: 1133–38.
Golla, W., H. Hofer, and M. L. East. 1999. Within-litter sibling aggression in spotted hyaenas: Effect of maternal nursing, sex and age. *Animal Behaviour* 58: 715–726.
Gottlander, K. 1987. Parental feeding behaviour and sibling competition in the pied flycatcher *Ficedula hypoleuca. Ornis Scandinavica* 18: 269–276.
Gould, S. J. 1982. The guano ring. *Natural History* 91: 12–19.
Gould, S. J., and R. C. Lewontin. 1979. The spandrels of San Marco and the Panglossian paradigm: A critique of the adaptationist programme. *Proceedings of the Royal Society of London B* 205: 581–598.
Gray, E. M. 1997. Do female red-winged blackbirds benefit genetically from seeking extra-pair copulations? *Animal Behaviour* 53: 605–623.
Grbíc, M., P. J. Ode, and M. R. Strand. 1992. Sibling rivalry and brood sex ratios in polyembryonic wasps. *Nature* 360: 254–256.
Griffin, C. R., P. W. C. Paton, and T. S. Baskett. 1998. Breeding ecology and behavior of the Hawaiian Hawk. *Condor* 100: 654–662.
Gustafsson, L., and W. J. Sutherland. 1988. The costs of reproduction in the collared flycatcher *Ficedula hypoleuca. Nature* 335: 813–815.
Gwynne, D. T. 1981. Sexual difference theory: Mormon crickets show role reversal in mate choice. *Science* 213: 779–780.
Gwynne, D. T., and L. W. Simmons. 1990. Experimental reversal of courtship in an insect. *Nature* 346: 172–174.
Hahn, D. C. 1981. Asynchronous hatching in the laughing gull: Cutting losses and reducing rivalry. *Animal Behaviour* 29: 421–427.
Haig, D. 1987. Kin conflict in seed plants. *Trends in Ecology and Evolution* 2: 337–340.
——— 1992. Brood reduction in gymnosperms. In M. A. Elgar and B. J. Crespi, eds., *Cannibalism,* 63–84. Oxford University Press.
Haig, D., and C. Graham. 1991. Genomic imprinting and the strange case of the insulin-like Growth Factor II receptor. *Cell* 64: 1045–46.

Haig, D., and M. Westoby. 1991. Genomic imprinting in endosperm: Its effect on seed development in crosses between species, and between different ploidies of the same species, and its implications for the evolution of apomixis. *Proceedings of the Royal Society of London* B 333: 1–13.

Haldane, J. B. S. 1955. Population genetics. *New Biology,* no. 18: 34–51.

Hamilton, W. D. 1964a. The genetical evolution of social behaviour. *Journal of Theoretical Biology* 7: 1–16.

——— 1964b. The genetical evolution of social behaviour. *Journal of Theoretical Biology* 7: 17–52.

——— 1967. Extraordinary sex ratios. *Science* 156: 477–488.

——— 1979. Wingless and fighting males in figwasps and other insects. In M. S. Blum and N. A. Blum, eds., *Sexual selection and reproductive competition in insects,* 167–220. Academic Press.

Hamilton, W. D., and M. Zuk. 1982. Heritable true fitness and bright birds: A role for parasites. *Science* 218: 384–387.

Hart, J. L. 1973. Pacific fishes of Canada. *Fisheries Research Board of Canada Bulletin* 180.

Harvey, J. A., L. S. Corley, and M. R. Strand. 2000. Competition induces adaptive shifts in caste ratios of a polyembryonic wasp. *Nature* 406: 183–186.

Haskell, D. 1994. Experimental evidence that nestling begging behaviour incurs a cost due to nest predation. *Proceedings of the Royal Society of London* B 257: 161–164.

——— 2002. Begging behaviour and nest predation. In J. Wright and M. L. Leonard, eds., *The evolution of begging,* 163–172. Kluwer Academic.

Hausfater, G., and S. B. Hrdy, eds. 1984. *Infanticide: Comparative and evolutionary perspectives.* Aldine.

Heaney, V., and P. Monaghan. 1996. Optimal allocation of effort between reproductive phases: The trade-off between incubation costs and subsequent brood rearing capacity. *Proceedings of the Royal Society of London B* 263: 1719–24.

Hébert, P. N., and R. M. R. Barclay. 1986. Asynchronous and synchronous hatching: Effect on early growth and survivorship of herring gull, *Larus argentatus,* chicks. *Canadian Journal of Zoology* 64: 2357–62.

Hébert, P. N., and S. G. Sealy. 1993. Hatching asynchrony and feeding rates in yellow warblers *Dendroica petechia:* A test of the sexual conflict hypothesis. *American Naturalist* 142: 881–892.

Högstedt, G. 1980. Evolution of clutch size in birds: Adaptive variation in relation to territory quality. *Science* 210: 1148–50.

Horsfall, J. 1984. Brood reduction and brood division in coots. *Animal Behaviour* 32: 216–225.

Houston, A. I., and N. B. Davies. 1985. The evolution of co-operation and life history in the dunnock *Prunella modularis*. In R. Sibly and R. Smith, eds., *Behavioural ecology: The ecological consequences of adaptive behaviour,* 471–487. Blackwell Scientific.

Howe, H. F. 1976. Egg size, hatching asynchrony, sex, and brood reduction in the common grackle. *Ecology* 57: 1195–1207.

Hrdy, S. B. 1977. *The langurs of Abu: Female and male strategies of reproduction.* Harvard University Press.

——— 1979. Infanticide among animals: A review, classification, and examination of the implications for the reproductive strategies of females. *Ethology and Sociobiology* 1: 13–40.

Huck, U. W. 1984. Infanticide and the evolution of pregnancy block in rodents. In G. Hausfater and S. B. Hrdy, eds., *Infanticide,* 349–360. Aldine.

Hussell, D. J. T. 1988. Supply and demand in tree swallow broods: A model of parent-offspring food provisioning interactions in birds. *American Naturalist* 131: 175–202.

Ingram, C. 1959. The importance of juvenile cannibalism in the breeding biology of certain birds of prey. *Auk* 76: 218–226.

——— 1962. Cannibalism by nestling short-eared owls. *Auk* 79: 715.

Inoue, Y. 1985. The process of asynchronous hatching and sibling competition in the little egret *(Egretta garzetta). Col. Waterbirds* 9: 1–12.

Jarvis, M. J. F. 1974. The ecological significance of clutch size in the South African gannet *(Sula capensus). Journal of Animal Ecology* 43: 1–17.

Jenni, D. A. 1969. A study of the ecology of four species of herons during the breeding season at Lake Alice, Alachua County, Florida. *Ecological Monograph* 39: 245–270.

Johnston, K., J. C. Bednarz, and S. Zack. 1987. Crested penguins: Why are first eggs smaller? *Oikos* 49: 347–349.

Johnstone, R. A. 1997. The evolution of animal signals. In J. R. Krebs and N. B. Davies, eds., *Behavioural ecology: An evolutionary approach,* 4th ed., 155–178. Sinauer.

Johnstone, R. A., and H. C. J. Godfray. 2002. Models of begging as a signal of need. In J. Wright and M. L. Leonard, eds., *Begging in birds,* 1–20. Kluwer Academic.

Kacelnik, A., P. A. Cotton, L. Stirling, and J. Wright. 1995. Food allocation among nestling starlings: Sibling competition and the scope of parental choice. *Proceedings of the Royal Society of London* B 259: 259–263.

Kahl, M. P. 1964. Food ecology of the wood stork (*Mycteria americana*) in Florida. *Ecological Monograph* 34: 97–117.

Kilner, R. 1997. Mouth colour is a reliable signal of need in begging canary nestlings. *Proceedings of the Royal Society of London B* 264: 963–968.

Kilner, R., and R. A. Johnstone. 1997. Begging the question: Are offspring solicitation behaviours signals of need? *Trends in Ecology and Evolution* 12: 11–15.

Kilner, R., D. G. Noble, and N. B. Davies. 1999. Signals of need in parent-offspring communication and their exploitation by the cuckoo, *Cuculus canorus*. *Nature* 397: 667–672.

Kłosowski, G., T. Kłosowski, and P. Zieliński. 2002. A case of parental infanticide in the black stork *Ciconia nigra*. *Avian Science* 2: 59–62.

Koenig, W. D., and R. L. Mumme. 1987. *Population ecology of the cooperatively breeding acorn woodpecker.* Princeton University Press.

Komdeur, J., S. Daan, J. Tinbergen, and C. Mateman. 1997. Extreme adaptive modification in sex ratio of the Seychelles warbler's eggs. *Nature* 385: 522–525.

Kozlowski, J., and S. C. Stearns. 1989. Hypotheses for the production of excess zygotes: Models of bet-hedging and selective abortion. *Evolution* 43: 1369–77.

Lack, D. 1947. The significance of clutch-size. Parts 1 and 2. *Ibis* 89: 302–352.

——— 1948. The significance of clutch-size. Part 3. *Ibis* 90: 25–45.

——— 1954. *The natural regulation of animal numbers*. Clarendon Press.

——— 1966. *Population studies in birds*. Clarendon Press.

——— 1968. *Ecological adaptations for breeding in birds*. Methuen.

Lamey, T. C. 1990. Hatch asynchrony and brood reduction in penguins. In L. S. Davis and J. Darby, eds., *Penguin biology*, 399–417. Academic Press.

——— 1992. Egg-size differences, hatch asynchrony, and obligate brood reduction in crested penguins. Ph.D. diss., University of Oklahoma, Norman.

——— 1993. Territorial aggression, timing of egg loss, and egg-size differences in rockhopper penguins, *Eudyptes c. chrysocome*, on New Island, Falkland Islands. *Oikos* 66: 293–297.

Lamey, T. C., R. M. Evans, and J. D. Hunt. 1996. Insurance reproductive value and facultative brood reduction. *Oikos* 77: 285–290.

Lamey, T. C., and C. S. Lamey. 1994. Hatch synchrony and bad food years. *American Naturalist* 143: 734–738.

Legge, S. 2000. Siblicide in the cooperatively breeding laughing kookaburra *(Dacelo novaeguineae)*. *Behavioral Ecology and Sociobiology* 48: 293–302.

Legge, S., and A. Cockburn. 2000. Social and mating system of cooperatively breeding laughing kookaburra *(Dacelo novaeguineae)*. *Behavioral Ecology and Sociobiology* 47: 220–229.

LeMasurier, A. D. 1987. A comparative study of the relationship between host size and brood size in *Apateles* spp. (Hymenoptera: Braconidae). *Ecological Entomology* 12: 383–393.

Leonard, M. L., and A. G. Horn. 1996. Provisioning rules in tree swallows. *Behavioral Ecology and Sociobiology* 38: 341–347.

Lichtenstein, G. 1998. Parasitism by shiny cowbirds of rufous-bellied towhees. *Condor* 100: 680–687.

Lichtenstein, G., and S. G. Sealy. 1998. Nestling competition, rather than supernormal stimulus, explains the success of parasitic brown-headed cowbirds. *Proceedings of the Royal Society of London B* 265: 249–254.

Ligon, J. D., R. Thornhill, M. Zuk, and K. Johnston. 1990. Male-male competition, ornamentation and the role of testosterone in sexual selection in red jungle fowl. *Animal Behaviour* 40: 367–373.

Lloyd, J. E. 1986. Firefly communication and deception: "Oh, what a tangled web." In. R. W. Mitchell and N. S. Thompson, eds., *Deception: Perspectives on human and nonhuman deceit.* SUNY Press.

Lotem, A. 1993. Learning to recognize nestlings is maladaptive for cuckoo *Cuculus canorus* hosts. *Nature* 362: 743–745.

Lougheed, L. W., and D. J. Anderson. 1999. Parent blue-footed boobies suppress siblicidal behavior of offspring. *Behavioral Ecology and Sociobiology* 45: 11–18.

Lyon, B. E., J. M. Eadie, and L. D. Hamilton. 1994. Parental choice selects for ornamental plumage in American coot chicks. *Nature* 371: 240–243.

Macedo, R. H., M. Cariello, and L. Muniz. 2001. Context and frequency of infanticide in communally breeding guira cuckoos. *Condor* 103: 170–175.

Machmer, M. M., and R. C. Ydenberg. 1998. The relative roles of hunger and size asymmetry in sibling aggression between nestling ospreys, *Pandion haliaetus. Canadian Journal of Zoology* 76: 181–186.

McRae, S. B., P. J. Weatherhead, and R. Montgomerie. 1993. American robin nestlings compete by jockeying for position. *Behavioral Ecology and Sociobiology* 33: 101–106.

Magrath, R. 1989. Hatch asynchrony and reproductive success in the blackbird. *Nature* 339: 536–538.

Maynard Smith, J. 1977. Parental investment: A prospective analysis. *Animal Behaviour* 25: 1–9.

——— 1982. *Evolution and the theory of games.* Cambridge University Press.

Maynard Smith, J., and G. A. Parker. 1976. The logic of asymmetric contests. *Animal Behaviour* 24: 159–175.

Mayr, E. 1991. *One long argument.* Harvard University Press.

Meffe, G. K., and M. L. Crump. 1987. Possible growth and reproductive benefits of cannibalism in the mosquitofish. *American Naturalist* 129: 203–212.

Mills, J. A., J. W. Yarrall, and D. A. Mills. 1996. Causes and consequences of mate fidelity in red-billed gulls. In J. M. Black, ed., *Partnerships in birds,* 286–304. Oxford University Press.

Mock, D. W. 1984a. Infanticide, siblicide, and avian nestling mortality. In G. Hausfater and S. B. Hrdy, eds., *Infanticide,* 3–30. Aldine.

——— 1984b. Siblicidal aggression and resource monopolization in birds. *Science* 225: 731–733.
——— 1985. Siblicidal brood reduction: The prey-size hypothesis. *American Naturalist* 125: 327–343.
——— 1987. Siblicide, parent-offspring conflict, and unequal parental investment by egrets and herons. *Behavioral Ecology and Sociobiology* 20: 247–256.
Mock, D. W., H. Drummond, and C. H. Stinson. 1990. Avian siblicide. *American Scientist* 78: 438–449.
Mock, D. W., and L. S. Forbes. 1992. Parent-offspring conflict: A case of arrested development? *Trends in Ecology and Evolution* 7: 409–413.
——— 1995. The evolution of parental optimism. *Trends in Ecology and Evolution* 10: 130–134.
Mock, D. W., and T. C. Lamey. 1991. The role of brood size in regulating egret sibling aggression. *American Naturalist* 138: 1015–26.
Mock, D. W., T. C. Lamey, and B. J. Ploger. 1987. Proximate and ultimate roles of food amount in regulating egret sibling aggression. *Ecology* 68: 1760–72.
Mock, D. W., T. C. Lamey, C. F. Williams, and A. Pelletier. 1987. Flexibility in the development of heron sibling aggression: An intraspecific test of the prey-size hypothesis. *Animal Behaviour* 35: 1386–93.
Mock, D. W., and G. A. Parker. 1986. Advantages and disadvantages of ardeid brood reduction. *Evolution* 40: 459–470.
——— 1997. *The evolution of sibling rivalry.* Oxford University Press.
——— 1998. Siblicide, family conflict, and the evolutionary limits of selfishness. *Animal Behaviour* 56: 1–10.
Mock, D. W., and B. J. Ploger. 1987. Parental manipulation of optimal hatch asynchrony in cattle egrets: An experimental study. *Animal Behaviour* 35: 150–160.
Monaghan, P., R. G. Nager, and D. C. Houston. 1998. The price of eggs: Increased investment in egg production reduces the offspring rearing capacity of parents. *Proceedings of the Royal Society of London B* 265: 1–5.
Mueller, U. 1991. Haplodiploidy and the evolution of facultative sex ratios in a primitively eusocial bee. *Science* 254: 442–444.
Mumme, R., W. Koenig, and F. Pitelka. 1983. Reproductive competition in the communal acorn woodpecker: Sisters destroy each others' eggs. *Nature* 306: 583–584.
Nakamura, K., and D. L. Kramer. 1982. Is sperm cheap? Limited male fertility and female choice in the lemon tetra (Pisces, Characidae). *Science* 216: 753–755.
Nathan, A., S. Legge, and A. Cockburn. 2001. Nestling aggression in broods of a siblicidal kingfisher, the laughing kookaburra. *Behavioral Ecology* 12: 716–725.

Nelson, J. B. 1964. Factors affecting clutch-size and chick growth in the North Atlantic gannet *Sula bassana. Ibis* 106: 63–77.

Nonacs, P. 1986. Ant reproductive strategies and sex allocation theory. *Quarterly Review of Biology* 61: 1–21.

Nuechterlein, G. L. 1981. Asynchronous hatching and sibling competition in western grebes. *Canadian Journal of Zoology* 59: 994–998.

Nuechterlein, G. L., and A. Johnson. 1981. The downy young of the hooded grebe. *Living Bird* 19: 69–71.

O'Connor, R. J. 1978. Brood reduction in birds: Selection for infanticide, fratricide, and suicide? *Animal Behaviour* 26: 79–96.

O'Gara, B. 1969. Unique aspects of reproduction in the female pronghorn (*Antilocapra americana*). *American Journal of Anatomy* 125: 217–232.

Olsen, P. D., and A. Cockburn. 1991. Female-biased sex allocation in peregrine falcons and other raptors. *Behavioral Ecology and Sociobiology* 28: 417–423.

Ono, M., T. Igarashi, E. Ohno, and M. Sasaki. 1995. Unusual thermal defence by a honeybee against mass attack by hornets. *Nature* 377: 334–336.

Osorno, J.-L., and H. Drummond. 1995. The function of hatching asynchrony in the blue-footed booby. *Behavioral Ecology and Sociobiology* 37: 265–273.

Packer, C., and A. E. Pusey. 1983a. Male takeovers and female reproductive parameters: A simulation of oestrous synchrony in lions *(Panthera leo). Animal Behaviour* 31: 334–340.

——— 1983b. Adaptations of female lions to infanticide by incoming males. *American Naturalist* 121: 716–728.

Parker, G. A. 1985. Models of parent-offspring conflict. V. Effects of the behaviour of the two parents. *Animal Behaviour* 33: 519–533.

Parker, G. A., and M. Macnair. 1978. Models of parent-offspring conflict. I. Monogamy. *Animal Behaviour* 26: 97–111.

Parker, G. A., D. W. Mock, and T. C. Lamey. 1989. How selfish should stronger sibs be? *American Naturalist* 133: 846–868.

Parsons, J. 1971. Cannibalism in herring gulls. *British Birds* 64: 528–537.

Payne, R. B. 1997. Avian brood parasitism. In D. H. Clayton and J. Moore, eds., *Host-parasite evolution: General principles and avian models.* Oxford University Press.

Petrie, M. 1994. Improved growth and survival of offspring of peacocks with more elaborate trains. *Nature* 371: 585–586.

Petrie, M., T. Halliday, and C. Sanders. 1991. Peahens prefer peacocks with elaborate trains. *Animal Behaviour* 41: 323–332.

Pfennig, D. W. 1992. Polyphenism in spadefoot toad tadpoles as a locally adjusted evolutionarily stable strategy. *Evolution* 46: 1408–20.

Pfennig, D. W., and J. P. Collins. 1993. Kinship affects morphogenesis in cannibalistic salamanders. *Nature* 362: 836–838.

Pfennig, D. W., J. P. Collins, and R. E. Ziemba. 1999. A test of alternative hypotheses for kin recognition in cannibalistic tiger salamanders. *Behavioral Ecology* 10: 436–443.

Pfennig, D. W., S. G. Ho, and E. A. Hoffman. 1998. Pathogen transmission as a selective force against cannibalism. *Animal Behaviour* 55: 1255–61.

Pfennig, D. W., H. K. Reeve, and P. W. Sherman. 1993. Kin recognition and cannibalism in spadefoot toad tadpoles. *Animal Behaviour* 46: 87–94.

Pfennig, D. W., P. W. Sherman, and J. P. Collins. 1994. Kin recognition and cannibalism in polyphenic salamanders. *Behavioral Ecology* 5: 225–232.

Ploger, B. J. 1997. Does brood reduction provide nestling survivors with a food bonus? *Animal Behaviour* 54: 1063–76.

Ploger, B. J., and D. W. Mock. 1986. Role of sibling aggression in distribution of food to nestling cattle egrets *(Bubulcus ibis)*. *Auk* 103: 768–776.

Poole, A. 1982. Brood reduction in temperate and sub-tropical ospreys. *Oecologia* 53: 111–119.

Price, K., H. Harvey, and R. Ydenberg. 1996. Begging tactics of nestling yellow-headed blackbirds *(Xanthocephalus xanthocephalus)* in relation to need. *Animal Behaviour* 51: 421–435.

Price, K., and R. Ydenberg. 1995. Begging and provisioning in broods of asynchronously-hatched yellow-headed blackbird nestlings. *Behavioral Ecology and Sociobiology* 37: 201–208.

Queller, D. C. 1989. Inclusive fitness in a nutshell. In P. H. Harvey and L. Partridge, eds., *Oxford surveys in evolutionary biology,* vol. 6, 73–109. Oxford University Press.

Ratniecks, F. L. W. 1988. Reproductive harmony via mutual policing by workers in eusocial Hymenoptera. *American Naturalist* 132: 217–236.

Ratniecks, F. L. W., and P. K. Visscher. 1989. Worker policing in the honeybee. *Nature* 342: 796–797.

Rees, E. C., P. Lievesley, R. A. Pettifor, and C. Perrins. 1996. Mate fidelity in swans: An interspecific comparison. In J. M. Black, ed., *Partnerships in birds,* 118–137. Oxford University Press.

Reid, W. R. 1987. The cost of reproduction in the glaucous-winged gull. *Oecologia* 74: 458–467.

Reynolds, P. S. 1996. Brood reduction and siblicide in black-billed magpies (*Pica pica*). *Auk* 113: 189–199.

Reznick, D. A., H. Bryga, and J. A. Endler. 1990. Experimentally induced life-history evolution in a natural population. *Nature* 346: 357–359.

Ricklefs, R. 1965. Brood reduction in the curve-billed thrasher. *Condor* 67: 505–510.

——— 1990. *Ecology.* 3rd ed. W. H. Freeman.

Rodríguez-Gironés, M. 1996. Siblicide: The evolutionary blackmail. *American Naturalist* 148: 101–122.

Rohwer, S. 1978. Parent cannibalism of offspring and egg raiding as a courtship strategy. *American Naturalist* 112: 429–440.

Roulin, A. 2002. The sibling negotiation hypothesis. In J. Wright and M. L. Leonard, eds., *Begging in birds,* 107–126. Kluwer Academic.

Rowe, E. G. 1947. The breeding biology of *Aquila verreauxi* Lesson. *Ibis* 89: 576–606.

Royle, N. J., I. R. Hartley, and G. A. Parker. 2002. Begging for control: When are offspring solicitation behaviours honest? *Trends in Ecology and Evolution* 17: 434–440.

Sargent, R. C. 1992. Ecology of filial cannibalism in fish: Theoretical perspectives. In M. A. Elgar and B. J. Crespi, eds. *Cannibalism,* 38–62. Oxford University Press.

Sato, T. 1986. A parasitic catfish of mouthbreeding cichlid fishes in Lake Tanganyika. *Nature* 323: 58–59.

Schwagmeyer, P. L. 1979. The Bruce effect: An evaluation of male/female advantages. *American Naturalist* 114: 932–938.

Schwagmeyer, P. L., D. W. Mock, T. C. Lamey, C. S. Lamey, and M. D. Beecher. 1991. Effects of sibling contact on hatch timing in an asynchronously hatching bird. *Animal Behaviour* 41: 887–894.

Sealy, S. G. 1995. Burial of cowbird eggs by parasitized yellow warblers: An empirical and experimental study. *Animal Behaviour* 49: 877–889.

Searcy, K. B., and M. R. Macnair. 1993. Developmental selection in response to environmental conditions of the maternal parent in *Mumulus guttatus. Evolution* 47: 13–24.

Searcy, W. A., and K. Yasukawa. 1995. *Polygyny and sexual selection in red-winged blackbirds.* Princeton University Press.

Seger, J. 1991. Cooperation and conflict in social insects. In J. R. Krebs and N. B. Davies, eds., *Behavioural ecology: An evolutionary approach,* 3rd ed., 338–373. Blackwell Scientific.

Sherman, P. W. 1977. Nepotism and the evolution of alarm calls. *Science* 197: 1246–53.

Sherman, P. W., H. K. Reeve, and D. W. Pfennig. 1997. Recognition systems. In J. R. Krebs and N. B. Davies, eds., *Behavioural ecology: An evolutionary approach,* 4th ed., 69–96. Blackwell Scientific.

Simmons, R. E. 1988. Offspring quality and the evolution of Cainism. *Ibis* 130: 339–357.

——— 1997. Why don't all siblicidal eagles lay insurance eggs? The egg quality hypothesis. *Behavioral Ecology* 8: 544–550.

Slagsvold, T., J. Sandvik, G. Rofstad, O. Lorentsen, and M. Husby. 1984. On the adaptive value of intraclutch egg-size variation in birds. *Auk* 101: 685–697.

Smale, L., K. E. Holekamp, M. Weldele, L. G. Frank, and S. E. Glickman. 1995. Competition and cooperation between litter-mates in the spotted hyaena, *Crocuta crocuta*. *Animal Behaviour* 50: 671–682.

Smith, C. C., and S. D. Fretwell. 1974. The optimal balance between size and number of offspring. *American Naturalist* 108: 499–506.

Smith, H. G., and R. Montgomerie. 1991. Nestling American robins compete with siblings by begging. *Behavioral Ecology and Sociobiology* 29: 307–312.

Smith, R. L. 1979a. Paternity assurance and altered roles in the mating behaviour of a giant water bug, *Abedus herberti* (Heteroptera, Belostomatidae). *Animal Behaviour* 27: 716–723.

——— 1979b. Repeated copulation and sperm precedence: paternity assurance for a male brooding water bug. *Science* 205: 1029–1031.

Spear, L. B., and N. Nur. 1994. Brood size, hatching order and hatching date: Effects on four life-history stages from hatching to recruitment in western gulls. *Journal of Animal Ecology* 63: 283–298.

Springer, S. 1948. Oviphagous embryos of the sand shark, *Carcharias taurus*. *Copeia* 1948: 153–157.

St. Clair, C. C. 1992. Incubation behavior, brood patch formation and obligate brood reduction in Fiordland crested penguins. *Behavioral Ecology and Sociobiology* 31: 409–416.

——— 1996. Multiple mechanisms of reversed hatching asynchrony in rockhopper penguins. *Journal of Animal Ecology* 65: 485–494.

St. Clair, C. C., J. R. Waas, R. C. St. Clair, and P. T. Boag. 1995. Unfit mothers? Maternal infanticide in royal penguins. *Animal Behaviour* 50: 1177–85.

Stamps, J. 1993. Begging in birds. *Etología* 3: 69–77.

Stamps, J., A. B. Clark, P. Arrowood, and B. Kus. 1989. Begging behaviour in budgerigars. *Ethology* 81: 177–192.

Stamps, J., and R. A. Metcalf. 1980. Parent-offspring conflict. In G. Barlow and J. Silverberg, eds., *Sociobiology: Beyond nature-nurture?* 598–618. Westview Press.

Stamps, J., R. A. Metcalf, and V. V. Krishnan. 1978. A genetic analysis of parent-offspring conflict. *Behavioral Ecology and Sociobiology* 3: 369–392.

Stanback, M. T., and W. D. Koenig. 1992. Cannibalism in birds. In M. A. Elgar and B. J. Crespi, eds., *Cannibalism*, 277–298. Oxford University Press.

Stearns, S. C. 1987. The selection arena hypothesis. In S. C. Stearns, ed., *The evolution of sex and its consequences*, 337–349. Birkhauser.

Stephenson, A. G. 1981. Flower and fruit abortion: Proximate causes and ultimate functions. *Annual Review of Ecology and Systematics* 12: 253–280.

Stephenson, A. G., and J. A. Winsor. 1986. *Lotus corniculatus* regulates offspring quality through selective fruit abortion. *Evolution* 40: 453–458.

Stinson, C. H. 1979. On the selective advantage of fratricide in raptors. *Evolution* 33: 1219–25.

Stoleson, S. H., and S. R. Beissinger. 1995. Hatching asynchrony and the onset of incubation in birds, revisited: When is the critical period? *Current Ornithology* 12: 191–270.

Strassmann, J. E., C. R. Hughes, D. C. Queller, S. Turillazzi, R. Cervo, S. K. Davis, and K. F. Goodnight. 1989. Genetic relatedness in primitively eusocial wasps. *Nature* 342:268–269.

Sutton, G. M. 1967. *Oklahoma birds.* University of Oklahoma Press.

Tait, D. E. N. 1980. Abandonment as a reproductive tactic in grizzly bears. *American Naturalist* 115: 800–808.

Teicher, M. H., and E. M. Blass. 1976. Suckling in newborn rats: Eliminated by nipple lavage, reinstated by pup saliva. *Science* 193: 422–424.

——— 1977. First suckling response of the newborn albino rat: The roles of olfaction and amniotic fluid. *Science* 198: 635–636.

Temme, D., and E. L. Charnov. 1987. Brood size adjustment in birds: Economical tracking in a temporally varying environment. *Journal of Theoretical Biology* 126: 137–147.

Thomas, F., Y. Kayser, and H. Hafner. 1999. Nestling size rank in Little Egrets *(Egretta garzetta)* influences subsequent breeding success of offspring. *Behavioral Ecology and Sociobiology* 45: 466–470.

Thornhill, R. 1976. Sexual selection and nuptial feeding behavior in *Bittacus apicalis* (Insecta: Mecoptera). *American Naturalist* 119: 529–548.

Thornhill, R., and J. Alcock. 1983. The evolution of insect mating systems. Harvard University Press.

Trillmich, F. 1986. Maternal investment and sex-allocation in the Galapagos fur seal, *Arctocephalus galapagoensis. Behavioral Ecology and Sociobiology* 19: 157–164.

Trivers, R. L. 1972. Parental investment and sexual selection. In B. Campbell, ed., *Sexual selection and the descent of man, 1871–1971,* 136–179. Aldine Atherton.

——— 1974. Parent-offspring conflict. *American Zoologist* 14: 249–264.

Trivers, R. L., and H. Hare. 1976. Haplodiploidy and the evolution of social insects. *Science* 191: 249–263.

Tyndale-Biscoe, C. H., and M. B. Renfree. 1987. *Reproductive physiology of marsupials.* Cambridge University Press.

Urrutia, L. P., and H. Drummond. 1990. Brood reduction and parental infanticide in Heermann's gull. *Auk* 107: 772–774.

van Heezik, Y. M., and P. J. Seddon. 1996. Scramble feeding in jackass penguins: Within-brood food distribution and the maintenance of sibling asymmetries. *Animal Behaviour* 51: 1383–90.

Vehrencamp, S. L. 1977. Relative fecundity and parental effort in communally nesting anis, *Crotophaga sulcirostris. Science* 197: 403–405.

Vermeer, K. 1963. The breeding biology of the glaucous-winged gull *Larus glaucescens* on Mandarte Island. *Occasional Papers of the British Columbia Provincial Museum* 13: 1–104.

Vesey-Fitzgerald, D. 1957. The breeding of the white pelican *(Pelecanus onocrotalus)* in the Rukwa Valley, Tanganyika. *Bulletin of the British Ornithological Club* 77: 1279.

Vince, M. A. 1969. Embryonic communication, respiration, and the synchronization of hatching. In R. A. Hinde, ed., *Bird vocalization,* 233–260. Cambridge University Press.

Viñuela, J. 1999. Sibling aggression, hatching asynchrony, and nestling mortality in the black kite (*Milvus migrans*). *Behavioral Ecology and Sociobiology* 45: 33–45.

Visscher, P. K. 1996. Reproductive conflict in honey bees: A stalemate of worker egg-laying and policing. *Behavioral Ecology and Sociobiology* 39: 237–244.

Voltura, K. M. 1998. Parental investment and offspring sex ratios in house sparrows, *Passer domesticus,* and cattle egrets, *Bubulcus ibis.* Ph.D. diss., University of Oklahoma, Norman.

vom Saal, F. S. 1984. Proximate and ultimate causes of infanticide and parental behavior in male house mice. In G. Hausfater and S. B. Hrdy, eds., *Infanticide,* 401–424. Aldine.

vom Saal, F. S., and F. H. Bronson. 1980. Sexual characteristics of adult female mice are correlated with their blood testosterone levels during prenatal development. *Science* 208: 597–599.

von Haartman, L. 1953. Was reizt den Trauerfliegeschnapper *(Muscicapa hypoleuca)* zu futtern? *Vogelwarte* 16: 157–164.

Waldman, B. 1988. The ecology of kin recognition. *Annual Review of Ecology and Systematics* 19: 543–571.

Waldman, B., and K. Adler. 1979. Toad tadpoles associate preferentially with siblings. *Nature* 282: 611–613.

Weatherhead, P. J. 1989. Sex ratios and host-specific reproductive success and impact of brown-headed cowbirds. *Auk* 106: 358–366.

Weir, B. J. 1971a. The reproductive physiology of the plains viscacha, *Lagostomus maximus. Journal of Reproduction and Fertility* 25: 355–364.

——— 1971b. The reproductive organs of the female plains viscacha, *Lagostomus maximus. Journal of Reproduction and Fertility* 25: 365–373.

Werschkul, D. F. 1979. Nestling mortality and the adaptive significance of early locomotion in the little blue heron. *Auk* 96: 116–130.

Westneat, D. F. 1993. Polygyny and extra-pair fertilizations in eastern red-winged blackbirds *(Agelaius phoeniceus). Behavioral Ecology* 4: 49–60.

Wiebe, K. L., and G. D. Bortolotti. 1994a. Food supply and hatching spans of birds: Energy constraints or facultative manipulations? *Ecology* 75: 813–823.

——— 1994b. Energetic efficiency of reproduction: The benefits of asynchronous hatching for American kestrels. *Journal of Animal Ecology* 63: 551–560.

Williams, G. C. 1966. Natural selection, the costs of reproduction, and a refinement of Lack's principle. *American Naturalist* 100: 687–690.

Williams, T. D. 1994. Intraspecific variation in egg size and egg composition in birds: Effects on offspring fitness. *Biology Reviews* 68: 35–59.

Willson, M. F., and N. Burley. 1983. *Mate choice in plants.* Princeton University Press.

Winkler, D. W. 1987. A general model for parental care. *American Naturalist* 130: 526–543.

Winkler, D. W., and K. Wallin. 1987. Offspring size and number: A life history model linking effort per offspring and total effort. *American Naturalist* 129: 708–720.

Withers, G. S., S. E. Fahrbach, and G. E. Robinson. 1995. Effects of experience and juvenile hormone on the organization of the mushroom bodies of honeybees. *Journal of Neurobiology* 26: 130–144.

Wright, J., and I. Cuthill. 1989. Manipulation of sex differences in parental care. *Behavioral Ecology and Sociobiology* 25: 171–181.

Wright, J., and M. L. Leonard, eds. 2002. *Begging in birds.* Kluwer Academic.

Wynne-Edwards, V. C. 1962. *Animal dispersion in relation to social behaviour.* Oliver and Boyd.

Ydenberg, R. C., and D. E. Bertram. 1989. Lack's clutch size hypothesis and brood enlargement on colonial seabirds. *Colonial Waterbirds* 12: 134–137.

Young, E. C. 1963. The breeding behaviour of the South Polar skua, *Catharacta maccormicki. Ibis* 105: 203–233.

Zuk, M., R. Thornhill, J. D. Ligon, and K. Johnston. 1990. Parasites and mate choice in red jungle fowl. *American Zoologist* 30: 235–244.

Acknowledgments

This book was written mainly during the spring of 1999, thanks to the willingness of the University of Oklahoma to let me cancel my teaching obligations for the term and focus on putting the first draft together. I have heard from no students mourning the loss of that statistics class, but if such exist I apologize to them. I finished the book when a sabbatical provided a second block of time in early 2003. For reading and offering comments on drafts of anywhere from one to many chapters, I thank Jennifer Alig, Gabriela Lichtenstein, Steve Mock, Johnnie Dennis, Colleen St. Clair, Scott Forbes, and Beth Clifford. For reading the whole book and offering both encouragement and valuable suggestions, I thank John Alcock, Hugh Drummond, Sarah Legge, Sarah Stai, and my diligent and impressively patient editors, Michael Fisher and Camille Smith. I also thank Dave Pfennig, Fritz Trillmich, Bruce Lyon, Sarah Legge, Gary Nuechterlein, David Fraser, Mike Collopy, Carl Marti, Colleen St. Clair, Scott Forbes, M. Strange, Noel Snyder, Piotr Zielinski, and Grzegorz and Tomasz Kłosowski for use of their rare and excellent photographs.

Index